T0282394

Cambridge Elements ≡

Elements in the Structure and Dynamics of Complex Networks
edited by
Guido Caldarelli
Ca' Foscari University of Venice

RECONSTRUCTING NETWORKS

Giulio Cimini
University of Rome Tor Vergata

Rossana Mastrandrea
IMT School for Advanced Studies

Tiziano Squartini
IMT School for Advanced Studies

CAMBRIDGE
UNIVERSITY PRESS

University Printing House, Cambridge CB2 8BS, United Kingdom

One Liberty Plaza, 20th Floor, New York, NY 10006, USA

477 Williamstown Road, Port Melbourne, VIC 3207, Australia

314–321, 3rd Floor, Plot 3, Splendor Forum, Jasola District Centre,
New Delhi – 110025, India

103 Penang Road, #05–06/07, Visioncrest Commercial, Singapore 238467

Cambridge University Press is part of the University of Cambridge.

It furthers the University's mission by disseminating knowledge in the pursuit of
education, learning, and research at the highest international levels of excellence.

www.cambridge.org
Information on this title: www.cambridge.org/9781108726818
DOI: 10.1017/9781108771030

© Giulio Cimini, Rossana Mastrandrea, and Tiziano Squartini 2021

First published 2021

A catalogue record for this publication is available from the British Library.

ISBN 978-1-108-72681-8 Paperback
ISSN 2516-5763 (online)
ISSN 2516-5755 (print)

Reconstructing Networks

Elements in the Structure and Dynamics of Complex Networks

DOI: 10.1017/9781108771030
First published online: August 2021

Giulio Cimini
University of Rome Tor Vergata
Rossana Mastrandrea
IMT School for Advanced Studies, Lucca
Tiziano Squartini
IMT School for Advanced Studies, Lucca
Author for correspondence: Tiziano Squartini, tiziano.squartini@imtlucca.it

Abstract: Complex networks datasets often come with the problem of missing information: interactions data that have not been measured or discovered, may be affected by errors, or are simply hidden because of privacy issues. This Element provides an overview of the ideas, methods and techniques to deal with this problem and that together define the field of network reconstruction. Given the extent of the subject, the authors focus on the inference methods rooted in statistical physics and information theory. The discussion is organized according to the different scales of the reconstruction task, that is, whether the goal is to reconstruct the macroscopic structure of the network, to infer its mesoscale properties, or to predict the individual microscopic connections.

Keywords: network reconstruction, maximum-entropy inference, exponential random graphs, mesoscale structures, link prediction

ISBNs: 9781108726818 (PB), 9781108771030 (OC)
ISSNs: 2516-5763 (online), 2516-5755 (print)

Contents

1 Introduction

Missing information: a general problem. Laying at the heart of the scientific method, data analysis is about using data to validate models, acquire useful information and support decision-making. When the data is incomplete, so are the conclusions that can be drawn from it. Unfortunately, the problem of missing data is a common occurrence, both in science and in many practical situations. Even in the era of Big Data, data can be incomplete for a variety of reasons – such as sheer lack of information, annotation errors, collection problems and privacy concerns. The problem is even more severe when the data has a non-homogeneous structure, because it describes non-trivial, systemic interconnection patterns, such as those characterizing complex networks. To be more concrete, we draw a few examples.

Consider a system biologist looking for the proteins in an organism that have physical, or functional, pairwise interactions. The scientist would need to pick two candidate proteins and set up an experiment to determine whether they interact or not. Blindly considering all possible pairs is unfeasible because experiments can be quite costly: this is why interactions within the proteome are largely unknown, whence the need to pick good candidates for the experiment, using prior information on those interactions that have already been discovered (Redner, 2008; Guimerà and Sales-Pardo, 2009). As another example, consider a social scientist trying to build up a given social network. Two types of problems can arise in this context: i) the available data reports only aggregated statistics on individuals (e.g. the total number of contacts) without disclosing sensible information such as the identities of friends; and ii) the network is extremely large to be explored by crawling algorithms, whence the need to consider subsamples that are representative (Leskovec and Faloutsos, 2006; Liben-Nowell and Kleinberg, 2007).

Farther from the classical typical scientific domain, consider an entrepreneur running an e-commerce platform that sells books. In order to improve sales, it would be a good idea to set up a *recommender system* that shows customers the books they may be interested in buying. The algorithm works well if it is able to predict customer tastes (i.e. possible future purchases) using their buying records (Lü et al., 2012). As a final example we take a regulator working in a central bank. Her job is to run stress tests to determine whether a given bank can withstand a crisis event. Since in a financial system losses and distress propagate through the various financial exposures banks have with each other and with other financial institutions, to accomplish her task properly the regulator should know the detailed network of exposures (who is exposed with whom, and to what extent). Unfortunately this information is confidential, and the regulator must resort to publicly disclosed information, i.e. the balance sheet of the banks containing only their aggregate exposures (Squartini et al., 2018; Anand et al., 2018).

The common theme of all these situations is that the system at hand is a *network*, namely a system that independently of its nature can be modelled by a complex pattern of interactions (the *links*) between its constituents (the *nodes*). When the network is known only partially, the task is to reconstruct the unknown part. The techniques constituting the field known as *network reconstruction* precisely aim at inferring the (unknown) structure of a network, making an optimal use of the partial knowledge about its properties (Squartini et al., 2018; Lü and Zhou, 2011).

Approaching network reconstruction. Generally speaking, the fundamental assumption at the basis of network reconstruction is *statistical homogeneity*: the empirically observed network structures should be representative of the statistical properties concerning the network as a whole. The validity of such an assumption is the necessary condition for a reconstruction algorithm to work. Clearly, this approach limits the accuracy that can be achieved when reconstructing strongly heterogeneous structures. However, it prevents possible inference biases introduced by arbitrary assumptions not supported by the available information.

In order to deal with the problem of missing information, many different approaches have been attempted so far. Among the most successful ones there are those defined within the framework of information theory (Cover and Thomas, 2006). In a nutshell, these methods prescribe to 1) consider all configurations that are compatible with the available information (an *ensemble*, in the jargon of statistical mechanics) and 2) assign a degree of plausibility to each of them. As it has been proven elsewhere, the least-biased way to do this rests upon the renowned *entropy maximization* prescription (Cover and Thomas, 2006; Jaynes, 1957). Notably, this approach naturally leads to the Exponential Random Graphs (ERG) formalism (Park and Newman, 2004b; Cimini et al., 2019). The importance of ERG models within the network reconstruction field is motivated by three desirable features they possess: *analytical character*, *general applicability* and *versatility*. This is why a large portion of this Element is devoted to discussing the applications of such a powerful formalism.

A quick overview of the Element. The discussion of the network reconstruction problem is divided into three sections, according to the scale of the reconstruction task: *macroscale*, *mesoscale* and *microscale*. This distinction is intended to provide a wide overview of the reconstruction techniques while presenting detailed results in some specific contexts.

The section **Network reconstruction at the macroscale** focuses on the inference of global features of the network, such as *(dis)assortative* and *hierarchical* patterns. In this case, reconstruction techniques are typically informed on node-specific properties (and possibly on trends that characterize the network as a whole), without considering any specific topological detail (i.e. the occurrence of a particular link). After describing the general ideas and results, particularly in the context of ERG, we will delve into the estimation of *systemic risk* in a partially-accessible network. As already mentioned, this exercise is particularly relevant for financial networks, where the knowledge of the interconnections between financial institutions is required to run stress tests and assess the stability of the system.

The section **Network reconstruction at the mesoscale** instead deals with the detection and reproduction of network patterns like *modular, core-periphery* and *bipartite* structures. The topic is of great interest for disciplines as diverse as epidemiology, finance, biology and sociology as it ultimately boils down to identifying some sort of *structural* or *functional* similarity between nodes. The presence of mesoscale patterns then affects a wide range of dynamical processes *on* networks (e.g. information and epidemic spreading, fake-news diffusion, etc.) whence the need to properly account for them. A fundamental point is to understand to what extent *accessible* node properties are informative about the presence of mesoscale structures.

Finally, the section **Network reconstruction at the microscale** is devoted to the topic of *single link* inference, a problem that is better known as *link prediction*. In contrast to the network reconstruction problem at the macro- and at the meso-scale, when considering the micro-scale many details of the network are known (typically a large number of connections) and the goal is to predict those links that are either not known because the source data used to define the network is incomplete, or simply do not exist yet. We will review the link prediction techniques that build on the partial knowledge on the network, and not on any additional information like nodes features.

2 Network Reconstruction at the Macroscale

A network is defined as a set of constituent elements (the *nodes*) and a set of connections (the *links*) among them. Mathematically speaking, a network is a graph with nontrivial topological features. In practice, networks are the natural way to represent and model a large class of very diverse systems, and thus we can speak of technological and information networks, social and economic networks as well as biological and brain networks.

Macroscale Properties: An Overview

Binary Properties

Let us start by introducing the basic notation and the macroscale properties of *binary, undirected* (*directed*) *graphs* with N nodes. Graphs of this kind are completely specified by a symmetric (generally asymmetric) $N \times N$ *adjacency matrix* $\mathbf{A}$, whose generic entry is either $a_{ij} = 0$ or $a_{ij} = 1$, respectively indicating the absence or the presence of a connection between nodes i and j (from i to j). As usual, *self-loops*, namely links starting and ending at the same node, will be ignored (in formulas, $a_{ii} = 0, \forall i$). The description above also applies to *bipartite* graphs, where nodes form two disjoint sets that are not connected internally.

Connectance. The simplest macroscopic characterization of a network is the connectance, or *link density*, defined as

$$\rho(\mathbf{A}) = \frac{2L}{N(N-1)}, \tag{2.1}$$

where $L(\mathbf{A}) = \sum_{i<j} a_{ij} \equiv L$ is the total number of links in the network. Thus $\rho(\mathbf{A})$ is the fraction of node pairs that are connected by a link. For directed networks, $\rho(\mathbf{A}) = \frac{L}{N(N-1)}$ with $L = \sum_{i \neq j} a_{ij}$.

Notably, real-world networks are usually characterized by a very low density of links (i.e. they are sparse). Reproducing the network connectance is a sort of baseline requirement of any reconstruction method. The simplest model satisfying this requirement is the *Erdös-Renyi* (ER) random graph (Erdös and Rényi, 1960; Park and Newman, 2004b). According to this model, the probability p_{ij} of a connection between nodes i and j (that is, the average value of the adjacency matrix element a_{ij} in the model, $\langle a_{ij} \rangle_{\text{ER}}$) reads

$$p_{ij}^{\text{ER}} = p = \frac{2L}{N(N-1)} = \rho, \ \forall \ i < j; \tag{2.2}$$

therefore, any two nodes establish a connection with the same probability p.

Degrees. The *degree* of a node counts the number of its neighbors, or equivalently the number of its incident connections. In formulas, $k_i(\mathbf{A}) = \sum_{j \neq i} a_{ij}, \forall i$. An important and ubiquitous characterization of real-world networks is the *heavy-tailed* shape of the degree distribution, with a few *hub* nodes that are highly connected ($k_{hub} = O(N)$) and the vast majority of other nodes (of the order $O(N)$) with a small degree. Although the mathematical nature of these heavy-tailed distributions is still debated, often they have been found to be *scale-free* (Caldarelli, 2007; Barabási, 2009):

$$P(k) \sim k^{-\gamma}, \, 2 < \gamma < 3, \tag{2.3}$$

for which the "typical" degree is simply missing. In any event, the strong heterogeneity of the degree distribution is the basic feature that makes networks different from homogeneous systems and regular lattices. Therefore, any good reconstruction algorithm should be able to reproduce it.[1] Notice that such a requirement rules out the ER model as a potentially good reconstruction model: in fact, although it ensures that the link density is reproduced, it fails to preserve the degree heterogeneity, since the model average $\langle k_i \rangle_{ER} = \sum_{j \neq i} \langle a_{ij} \rangle_{ER} = \sum_{j \neq i} p_{ij}^{ER} = 2L/N$, $\forall \, i$. Such evidence has motivated the definition of the Chung-Lu (CL) model (Chung and Lu, 2002), according to which

$$p_{ij}^{CL} = \frac{k_i k_j}{2L}, \, \forall \, i \neq j; \tag{2.4}$$

by definition, then, $\langle k_i \rangle_{CL} = \sum_{j \neq i} \langle a_{ij} \rangle_{CL} = \sum_{j \neq i} p_{ij}^{CL} \simeq k_i$, $\forall \, i$.

In the directed case, there are two kinds of degree: the total number of links outgoing from a node (the *out-degree* $k_i^{out}(\mathbf{A}) = \sum_{j \neq i} a_{ij}$, $\forall \, i$) and the total number of links incoming to a node (the *in-degree* $k_i^{in}(\mathbf{A}) = \sum_{j \neq i} a_{ji}$, $\forall \, i$). The directed extension of the Chung-Lu model (DCL) reads

$$p_{ij}^{DCL} = \frac{k_i^{out} k_j^{in}}{L}, \, \forall \, i \neq j. \tag{2.5}$$

Assortativity. Generally speaking, this term indicates the tendency of nodes to establish connections with other nodes having either similar (*positive assortativity*) or different (*negative assortativity* or *disassortativity*) characteristics. Particularly relevant in the study of complex networks is the assortativity *by degree*. In this case, assortativity can be studied by considering the *average nearest neighbor degree* (ANND), which for generic node i is defined as

$$k_i^{nn}(\mathbf{A}) = \frac{\sum_{j \neq i} a_{ij} k_j}{k_i}, \, \forall \, i. \tag{2.6}$$

ANND is a quadratic function of the adjacency matrix and thus is a *second-order* network property. Plotting k_i^{nn} versus k_i reveals the two-point correlation structure of the network: an increasing trend corresponds to an assortative pattern (poorly connected nodes are connected to other poorly connected nodes, highly connected nodes are connected to other highly connected nodes), while a decreasing trend corresponds to the opposite disassortative pattern (poorly connected nodes are connected to highly connected nodes and vice versa). Notice that assortativity is typically observed in social networks (where it is also

---

[1] Moreover, preserving the degrees automatically ensures that the link density is preserved.

known by the term *homophily*), whereas economic and technological networks are usually disassortative (Newman, 2002).

Assortativity acts as the test bench for the CL model. Since $\langle k_i^{nn} \rangle_{\text{CL}} \simeq \frac{\sum_{j \neq i} p_{ij}^{\text{CL}} k_j}{k_i} = \frac{\sum_{j \neq i} k_j^2}{2L}$, $\forall$ i, in this model k_i^{nn} is weakly dependent on node i – basically, the ANND is the same for all nodes (Squartini and Garlaschelli, 2011). As a consequence, the CL model is not capable of reproducing any (dis)assortativity, thus it lacks one of the characteristic features of real-world networks. The solution lies in the definition of a more refined model, the *Configuration Model* (CM – see below).

When considering directed networks, the ANND can be generalized in five different ways (see Squartini et al., 2011a for further details).

Hierarchy. Assortativity and ANND account for second-order interactions, that is, interactions between nodes along patterns of length two. Third-order interactions (i.e. three-point correlations) are instead typically measured through the *clustering coefficient*, which for any node i is defined as the percentage of pairs of neighbors of i that are also neighbors of each other:

$$c_i(\mathbf{A}) = \frac{\sum_{j \neq i} \sum_{k \neq i,j} a_{ij} a_{ik} a_{jk}}{k_i(k_i - 1)}, \ \forall \ i; \tag{2.7}$$

otherwise stated, c_i measures the fraction of potential triangles attached to i (and defined by the product $a_{ij}a_{ik}$) that are actually realized (i.e. closed by the third link, a_{jk}). A decreasing trend of c_i as a function of k_i indicates that neighbors of highly connected nodes are poorly interconnected, whereas neighbors of poorly connected nodes are highly interconnected. This behavior characterizes a *hierarchical* network (i.e., a network of densely connected subgraphs that are poorly interconnected). In real-world networks, a scale-free degree distribution often coexists with a large value of the clustering coefficient (Albert and Barabási, 2002).

As for the assortativity, the CL model predicts a value for the clustering coefficient that is only weakly dependent on i, thus calling for a more refined model to reproduce empirical patterns of real-world networks.

Generalizations to directed networks also exist for third-order quantities. Besides five different definitions of the clustering coefficient (Squartini et al., 2011a), there are thirteen possible patterns involving three nodes and all possible connections between them: these quantities are called *motifs* and, as discussed in Section 3, have been proven to play a fundamental role in the self-organization of biological, ecological, and cellular networks. Certain *structures* have been, in fact, suggested to promote specific *functions* (Milo et al., 2002).

Higher-Order Patterns. The presence of higher-order patterns can be inspected using the powers of the network adjacency matrix $\mathbf{A}$. Indeed, the entry indexed by i and j of $\mathbf{A}^n$ (i.e., the nth power of $\mathbf{A}$) counts the number of paths of length n existing between i and j (or from i to j).

A very popular higher-order pattern is given by the *shortest path length*, a concept entering into the definition of the well-known *small-world effect* (Watts and Strogatz, 1998). Small-worldness refers to the evidence that, in many real-world networks, two (apparently) competing features coexist: a large clustering coefficient and a small average shortest path length. More quantitatively, the small-world phenomenon is characterized by an average shortest path length typical of random graphs

$$\bar{d} \simeq d_{\text{random}} \propto \ln N \qquad (2.8)$$

(i.e., growing "slowly" with the size of the system) and by an average clustering coefficient typical of regular lattices (i.e. independent of the system size), much larger than that of a random graph

$$\bar{c} \gg \bar{c}_{\text{random}} \propto N^{-1} \qquad (2.9)$$

where, in both expressions, the term "random" refers to the ER model.

Nestedness. A pattern that has recently attracted much attention is the *nestedness*. It quantifies how much the *biadjacency* matrix of a bipartite network can be rearranged to let a triangular structure emerge (Johnson et al., 2013; Mariani et al., 2019). Several measures have been defined to quantify the nestedness, among which the NODF (an acronym for "Nestedness metric based on Overlap and Decreasing Fill") that quantifies a matrix "triangularity" by measuring the overlap between rows and between columns (Almeida-Neto et al., 2008). Nestedness has been observed in ecological and economic systems alike. The classical example of nested ecological systems is given by the interactions between plants and pollinators, where nestedness emerges due to the presence of generalist pollinators (being attracted by all species of plants) coexisting with specialist pollinators (being attracted by only a small number of species of plants). Such a structure has been argued to promote the stability of the ecosystem (Bascompte et al., 2003). For what concerns economic systems, nestedness is observed in the structure of countries' exports: a few very diversified countries have a large export basket, while others only export some simple products. Interestingly, this pattern contradicts classical economic theories, according to which countries should specialize and export only those products in which they have a competitive advantage, and implying a block-diagonal biadjacency matrix instead of a nested one (Tacchella et al., 2012).

Centrality. The concept of *centrality* aims at quantifying the "importance" of a node in a network (Newman, 2018a). Besides *degree centrality*, the centrality given by the degree, other well-known measures are the *closeness centrality*, defined as

$$C_i(\mathbf{A}) = \frac{1}{\bar{d}_i}, \ \forall \ i \tag{2.10}$$

(i.e., as the reciprocal of the average topological distance of a node from the others), and the *betweenness centrality*, defined as

$$B_i(\mathbf{A}) = \sum_{j \neq i} \sum_{k \neq i,j} \frac{\sigma_{jk}(i)}{\sigma_{jk}}, \ \forall \ i, \tag{2.11}$$

where σ_{jk} is the total number of shortest paths from node j to node k, and $\sigma_{jk}(i)$ is the number of these paths passing through i.

Most of the proposed centrality measures are computable only on undirected networks. A notable exception is the *PageRank centrality* (Page et al., 1999), which can be computed by solving the iterative equation

$$P_i(\mathbf{A}) = \frac{1-\alpha}{N} + \alpha \sum_{j \neq i} \left(\frac{a_{ji}}{k_j^{out}} \right) P_j(\mathbf{A}), \ \forall \ i. \tag{2.12}$$

In general, it is very difficult to reconstruct the patterns of centrality of a network, unless these are strongly correlated with the degree centrality (Barucca et al., 2018).

Reciprocity. In the specific case of directed networks, it is of particular interest to measure the percentage of links having a counterpart pointing in the opposite direction. This quantity is known as *reciprocity* and reads

$$r(\mathbf{A}) = \frac{L^{\leftrightarrow}}{L} = \frac{\sum_i \sum_{j \neq i} a_{ij} a_{ji}}{\sum_i \sum_{j \neq i} \sum_i a_{ij}}; \tag{2.13}$$

remarkably, different classes of real-world networks are characterized by different values of reciprocity (Garlaschelli and Loffredo, 2004b). For instance, reciprocity is a distinguishing feature of financial networks, being associated with the level of "trust" between banks (Squartini et al., 2013a).

Spectral Properties. This term refers to the features of eigenvalues and eigenvectors of both the adjacency matrix $\mathbf{A}$ and the *Laplacian matrix* $\mathbf{L} = \mathbf{D} - \mathbf{A}$ of the network.[2] (Here $\mathbf{D}$ is the diagonal matrix whose generic entry reads

---

[2] The focus on undirected binary networks is justified by the ease of treating symmetric matrices, a characteristic ensuring that eigenvalues are real, for example.

$d_{ii} = k_i$, $\forall\, i$.) While Laplacian spectral properties provide information on macroscale network properties like the number of connected components (that matches the multiplicity of the zero eigenvalue of $\mathbf{L}$), spectral properties of $\mathbf{A}$ provide information on higher-order patterns like cycles (Estrada and Knight, 2015) as well as on dynamical properties of spreading processes (Bardoscia et al., 2017). Notice that the reconstruction of spectral properties of empirical networks is still a largely underexplored topic, although a first result in this sense is provided by the Silverstein theorem (Silverstein, 1994).

Weighted Properties

While binary networks are characterized by an adjacency matrix whose entries assume only the values 0 and 1, *weighted, undirected (directed)* graphs are specified by a symmetric (generally asymmetric) $N \times N$ matrix $\mathbf{W}$ whose generic entry $w_{ij} > \,= 0$ quantifies the intensity of the link connecting nodes i and j: In the most general case, w_{ij} is a real number; however, in many cases w_{ij} assumes integer values. Naturally, $\mathbf{A}$ and $\mathbf{W}$ are related by the position $a_{ij} = \Theta[w_{ij}]$, $\forall\, i,j$, simply stating that any positive weight between i and j carries the information that i and j are indeed connected.

Weight Distribution. When links are characterized by "magnitudes," the first step is to inspect the distribution of these magnitudes. When considering real-world networks, weight distributions are often found to be *fat-tailed.*

Strengths. The weighted analogue of the degree is the so-called *strength*. It is defined as $s_i(\mathbf{W}) = \sum_{j \neq i} w_{ij}$, $\forall\, i$ (i.e., as the sum of the weight of the links connected to node i). Similar to the case of degrees, strength distributions are often found to be *fat-tailed*. When directed networks are considered, one speaks of *out-strength* and *in-strength*, respectively defined as $s_i^{out}(\mathbf{W}) = \sum_{j \neq i} w_{ij}$, $\forall\, i$ and $s_i^{in}(\mathbf{W}) = \sum_{j \neq i} w_{ji}$, $\forall\, i$.

From a network reconstruction perspective, strengths play an important role, since they often represent the only kind of information available for the system under consideration. The typical example is that of financial networks, where only the total *assets* and *liabilities* of each bank (respectively the out- and in-strengths of the respective node) are accessible. This has motivated the definition of the weighted analogue of the Chung-Lu model, also known as the *MaxEnt* (ME) recipe. Its directed version, reading

$$\hat{w}_{ij}^{\text{ME}} = \frac{s_i^{out} s_j^{in}}{W}, \quad \forall\, i \neq j \tag{2.14}$$

(with $W = \sum_i s_i^{out} = \sum_i s_i^{in}$), is extensively used to estimate the magnitude of links in economic and financial networks (Mistrulli, 2011; Upper, 2011; Squartini et al., 2018).

Weighted Assortativity. The concept of assortativity can be easily extended to the weighted case. The weighted counterpart of the average nearest neighbors degree of node i is the *average nearest neighbor strength* (ANNS):

$$s_i^{nn}(\mathbf{W}) = \frac{\sum_{j \neq i} a_{ij} s_j}{k_i}, \; \forall \, i. \tag{2.15}$$

Analogous to the binary case, the correlation between strengths can be inspected by plotting s_i^{nn} versus s_i. Note that since $\langle a_{ij} \rangle_{\mathrm{ME}} = p_{ij}^{\mathrm{ME}} = \Theta[\hat{w}_{ij}^{\mathrm{ME}}]$, the weighted version of the CL model always generates a very densely connected network and, as a consequence, a value for the ANNS of node i that is weakly dependent on i itself, that is, $\langle s_i^{nn} \rangle_{\mathrm{ME}} \simeq \frac{\sum_{j \neq i} p_{ij}^{\mathrm{ME}} s_j}{\langle k_i \rangle_{\mathrm{ME}}} \simeq \frac{\sum_{j \neq i} s_j}{N-1} \simeq \frac{2W}{N-1}, \; \forall \, i$ (Squartini and Garlaschelli, 2011). The weighted CL model thus suffers from the same limitations affecting the binary CL model.[3]

Weighted Hierarchy. A *weighted clustering coefficient* (WCC) can be defined to capture the "intensity" of the triangles in which node i participates (Squartini et al., 2011b):

$$c_i^w(\mathbf{W}) = \frac{\sum_{j \neq i} \sum_{k \neq i,j} (w_{ij} w_{jk} w_{ki})^{1/3}}{k_i(k_i - 1)}. \tag{2.16}$$

Contrary to what is observed for the vast majority of binary networks, plotting c_i^w versus s_i reveals an increasing trend for many real-world networks, indicating that nodes with larger total activity participate in more "intense" triangles.

For extensions of ANNS and WCC to directed networks, see Squartini et al. (2011b).

Weighted Reciprocity. A weighted version of link reciprocity can be defined as

$$r^w(\mathbf{W}) = \frac{W^{\leftrightarrow}}{W} = \frac{\sum_i \sum_{j \neq i} \min[w_{ij}, w_{ji}]}{\sum_i \sum_{j \neq i} w_{ij}}, \tag{2.17}$$

a quantity whose numerator accounts for the "minimum exchange" between any two nodes (Squartini et al., 2013b).

---

[3] As we will see in what follows, the weighted counterpart of the CM, namely the *Weighted Configuration Model* (WCM), does not represent the solution to this problem. We will need to consider degrees and strengths together as in the *Enhanced Configuration Model* (ECM).

Higher-Order Patterns and Centrality. In contrast to the purely binary case, higher-order patterns in weighted networks are rarely inspected. An attempt to define *weighted motifs* has been done in Onnela et al., 2005, while *weighted centrality measures* have been defined in Opsahl et al., 2010.

Macroscale Reconstruction of Economic and Financial Networks: A Quick Historical Survey

The network reconstruction problem can be formulated as follows. Let us consider the most general case of a weighted, directed network, represented by an $N \times N$ asymmetric matrix $\mathbf{W}$ with positive real entries $w_{ij} \; \forall \; i, j$. When considering financial networks, the generic entry w_{ij} may represent the value of the exposure of i Towards j; in the case of economic networks, it may represent the value of exports from country i to country j.

In the well-studied case of economic and financial networks, the available information is represented by the out-strength and in-strength sequences, that is, $s_i^{out}(\mathbf{W}) = \sum_{j \neq i} w_{ij}, \; \forall \; i$ and $s_i^{in}(\mathbf{W}) = \sum_{j \neq i} w_{ji}, \; \forall \; i$: the network reconstruction goal, thus, becomes the estimation of the generic entry w_{ij} via the aforementioned, aggregate information (Squartini et al., 2018; Anand et al., 2018).

One of the first-ever proposed reconstruction algorithms was born with the aim of inferring the value of direct exposures between financial institutions and is known as the *MaxEnt* algorithm (Wells, 2004; Mistrulli, 2011). The method prescribes to maximize the entropic functional

$$S = -\sum_i \sum_j w_{ij} \ln w_{ij} \qquad (2.18)$$

under the constraints represented by the out- and in-strengths (respectively assets and liabilities, in the jargon of finance). The solution to the aforementioned constrained maximization problem reads

$$\hat{w}_{ij}^{ME} = \frac{s_i^{out} s_j^{in}}{W}, \; \forall \; i \neq j, \qquad (2.19)$$

where $W = \sum_i \sum_{j \neq i} w_{ij}$ denotes the total economic value of the system at hand. The recipe above is simple and allows the constraints to be satisfied: in fact, $\hat{s}_i^{out} = \sum_j \hat{w}_{ij}^{ME} = s_i^{out}$ and analogously for s_i^{in}. However, it suffers from two major drawbacks. Firstly, constraints are satisfied only if the summation index runs over all values $j = 1 \ldots N$, including the ones corresponding to the diagonal entries: the method needs self-loops to work. Second, the generated network topologies are unrealistically densely connected – in fact, no entry can be predicted to be zero, unless either $s_i^{out} = 0$ or $s_i^{in} = 0$ for some nodes. This is quite a problem for financial networks, since employing excessively dense configurations for running the stress tests typically leads to underestimating the systemic risk (Mistrulli, 2011). On the other hand, excessively sparse configurations lead to systemic risk overestimation (Anand et al., 2015). Nevertheless, the MaxEnt prescription provides quite accurate estimates of the *magnitude* of empirical weights (Squartini et al., 2017a; Mazzarisi and Lillo, 2017; Almog et al., 2019).

A first attempt to build more realistic configurations is represented by the *iterative proportional fitting* (IPF) algorithm (Deming and Stephan, 1940; Bacharach, 1965; Fienberg, 1970). This is a simple recipe to obtain a matrix that (1) lies at the "minimum distance" from the MaxEnt matrix $\hat{\mathbf{W}}^{\text{ME}}$, (2) satisfies the constraints represented by the available information, and (3) admits the presence of a set of zero entries (in the simplest case, the diagonal ones). The main drawback of the IPF, however, is that of requiring the knowledge of the position of zeros *in advance*, a piece of information that is (practically) never accessible.

The evidence that the outcome of systemic risk estimation strongly depends on the link density has pushed many researchers to devise a way to tune the density of the reconstructed network (Mastromatteo et al., 2012; Drehmann and Tarashev, 2013; Halaj and Kok, 2013; Cimini et al., 2015b). This is typically achieved though a free parameter in the algorithm, intended to enforce the desired density.[4] As an "extreme" example, let us consider the *minimum density algorithm* (Anand et al., 2015), intended to find the network structure with minimum link density that still satisfies the weighted constraints. Although its main limitation is that of overestimating the impact of shocks on the predicted configuration (intuitively, the few admitted links carry the maximum possible load, thus propagating the largest possible shocks), its combined use with the MaxEnt algorithm allows one to provide an upper and a lower bound to the systemic risk.

The Exponential Random Graphs Framework

We now introduce the popular framework of Exponential Random Graphs that will be used to carry out the network reconstruction at the macroscale. The idea is that of building a network model able to reproduce a bunch of chosen quantities (the *constraints*) while ensuring that everything else is kept as random as possible. The core quantity of the method is *Shannon entropy*, defined as

---

[4] The attractiveness of these methods lies in the possibility of generating different topologies still satisfying the (weighted) constraints: in fact, since the link density is *chosen* rather than *reproduced*, these methods are well suited for defining possible scenarios over which to run stress tests.

$$S = - \sum_{\mathbf{W} \in \mathcal{W}} P(\mathbf{W}) \ln P(\mathbf{W}), \tag{2.20}$$

where $P(\mathbf{W})$ is the probability distribution of the specific network configuration $\mathbf{W}$ output by the model – defined over the ensemble $\mathcal{W}$ of allowable configurations. The constrained maximization of S represents an inference procedure that has been proved to be maximally noncommittal with respect to the missing information (Jaynes, 1957): in other words, it allows one to derive a probability distribution that minimizes the number of unjustified assumptions about the unavailable data.

More quantitatively, it can be implemented by defining the Lagrangean function

$$\mathscr{L}[P] = S - \sum_{m=0}^{M} \theta_m \left(\sum_{\mathbf{W} \in \mathcal{W}} P(\mathbf{A}) C_m(\mathbf{W}) - C_m^* \right)$$

$$= S - \sum_{m=0}^{M} \theta_m \left(\langle C_m \rangle - C_m^* \right), \tag{2.21}$$

with θ_m representing the Lagrange multiplier associated to the m-th constraint, $C_m(\mathbf{W})$ representing the value of the m-th constraint, measured on the configuration $\mathbf{W}$, $\langle C_m \rangle$ representing its average value over the ensemble $\mathcal{W}$ and C_m^* being the desired value of the m-th constraint. $C_0(\mathbf{W}) = C_0^* = 1$ sums up the normalization condition. Upon solving the equation $\frac{\delta \mathscr{L}[P]}{\delta P(\mathbf{W})} = 0$ we find the expression $P(\mathbf{W}|\vec{\theta}) = e^{-1 - \vec{\theta} \cdot \vec{C}(\mathbf{W})}$ that can be further rewritten as

$$P(\mathbf{W}|\vec{\theta}) = \frac{e^{-H(\mathbf{W}, \vec{\theta})}}{Z(\vec{\theta})}, \tag{2.22}$$

a formula defining the *Exponential Random Graphs* (ERG) formalism in its full generality, with the quantity $H(\mathbf{W}, \vec{\theta}) = \vec{\theta} \cdot \vec{C}(\mathbf{W}) = \sum_{m=1}^{M} \theta_m C_m(\mathbf{W})$ usually called *Graph Hamiltonian* (Park and Newman, 2004b).

As $P(\mathbf{W}|\vec{\theta})$ depends on the vector of parameters $\vec{\theta}$, we need a recipe to estimate them. Such a recipe comes from the *likelihood maximization principle*, prescribing to solve the system of equations

$$\frac{\partial \mathcal{L}(\vec{\theta})}{\partial \vec{\theta}} = \frac{\partial \ln P(\mathbf{W}^*|\vec{\theta})}{\partial \vec{\theta}} = \vec{0} \tag{2.23}$$

with respect to the unknowns. Here $\mathbf{W}^*$ indicates an empirical network configuration that meets the values of the imposed constraints. Notice that substituting Eq. (2.22) into Eq. (2.23) and solving it leads to the system of equations $\langle C_m \rangle (\vec{\theta}) = C_m^*$, $\forall\, m$, guaranteeing that the expected values of the constraints match the imposed ones (Squartini and Garlaschelli, 2011).

The Configuration Model (CM). The most popular model of the ERG family is the *configuration model*, where the constraints imposed are the node degrees (Park and Newman, 2004b; Cimini et al., 2019). The Hamiltonian thus reads

$$H(\mathbf{A}, \vec{\theta}) = \sum_i \theta_i k_i(\mathbf{A}) \tag{2.24}$$

and the probability of a network in the ensemble is

$$P(\mathbf{A}) = \prod_{i<j} p_{ij}^{a_{i\alpha}} (1 - p_{ij})^{1-a_{i\alpha}}, \tag{2.25}$$

where $p_{ij} = \frac{x_i x_j}{1+x_i x_j}$ stands for the probability that a link exists between nodes i and j. The parameters x are numerically determined by solving the following system of equations, induced by the maximization of the corresponding likelihood function.

$$k_i(\mathbf{A}^*) = \sum_{j(\neq i)} \frac{x_i x_j}{1 + x_i x_j} = \langle k_i \rangle_{\text{CM}}, \ \forall \ i. \tag{2.26}$$

Constraining Linear vs. Nonlinear Quantities. The ERG framework allows considering as a constraint any function of the adjacency matrix of the network. In what follows, we will focus on ERG models defined by linear constraints (i.e. linear functions of the adjacency matrix elements) for at least two reasons. Firstly, as we will see, linear models perform remarkably well when employed to reconstruct several networks of interest. Secondly, they are not affected by the (theoretical and practical) limitations characterizing models with nonlinear constraints. Examples of the latter ones are provided by the models studied in Park and Newman, 2004a and Park and Newman, 2005, constraining the total number of *links* and *two-stars* and the total number of *links* and *triangles*, respectively. What the authors find for these models is the presence of *phases* and phase transitions, analogous to the ones that characterize classical disordered systems of spins. Although interesting from a purely theoretical perspective, these models cannot be easily used for reproducing the properties of real-world networks. Other examples are provided by models specifying the degree distribution and degree–degree correlations (and clustering, to some extent), which, however, can be approached mainly through numerical Monte Carlo sampling methods (Coolen et al., 2009; Annibale et al., 2009; Orsini et al., 2015).

The Best-Performing Reconstruction Method

We now show how to apply the ERG framework in the representative case of financial networks, where nodes' out- and in-strengths are the only information on the network. To overcome the problems affecting the ME recipe, one may be tempted to solve Eq. (2.22) by imposing the out- and in-strength sequences as constraints. This model, known as *Directed Weighted Configuration Model* (DWCM) (Squartini and Garlaschelli, 2011), is characterized by a geometric weight-specific distribution

$$q_{ij}^{\text{DWCM}}(w_{ij}) = (x_i y_j)^{w_{ij}}(1 - x_i y_j), \ \forall \ i \neq j, \tag{2.27}$$

whose unknowns x, y can be estimated via the constraints equations

$$s_i^{out}(\mathbf{W}^*) = \sum_{j \neq i} \frac{x_i y_j}{1 - x_i y_j} = \sum_{j \neq i} \langle w_{ij} \rangle_{\text{DWCM}} = \langle s_i^{out} \rangle_{\text{DWCM}}, \ \forall \ i \tag{2.28}$$

$$s_i^{in}(\mathbf{W}^*) = \sum_{j \neq i} \frac{x_j y_i}{1 - x_j y_i} = \sum_{j \neq i} \langle w_{ji} \rangle_{\text{DWCM}} = \langle s_i^{in} \rangle_{\text{DWCM}}, \ \forall \ i. \tag{2.29}$$

An alternative estimation procedure characterizes the *Maximum-Entropy Capital Asset Pricing Model* (MECAPM) (Di Gangi et al., 2018), which prescribes to equate the expression for the expected weights under the DWCM to the MaxEnt estimate, that is,

$$\langle w_{ij} \rangle_{\text{DWCM}} = \frac{x_i y_j}{1 - x_i y_j} = \hat{w}_{ij}^{\text{ME}}, \ \forall \ i \neq j. \tag{2.30}$$

Solving this system of equations then leads to recovery of the numerical value of the expressions $x_i y_j$, $\forall \ i \neq j$.

Unfortunately, both the DWCM and MECAPM generate very dense network configurations and thus, like the ME recipe, perform poorly in reproducing the topological structure of the network (Squartini et al., 2018; Mazzarisi and Lillo, 2017). This drawback can be solved only by *imposing* some kind of topological information, besides the weighted one represented by the sequences $\{s_i^{out}\}_{i=1}^N$ and $\{s_i^{in}\}_{i=1}^N$. The need of adding some topological information leads to the definition of two broad classes of algorithms: those simultaneously imposing binary and weighted constraints and those acting iteratively on ad-hoc topologies. Among the algorithms belonging to the first group, a special mention is deserved by the *Enhanced Configuration Model* (ECM)[5] (Mastrandrea et al., 2014a), defined (in the simpler undirected case) by the recipe

$$q_{ij}^{\text{ECM}}(w_{ij}) = \begin{cases} 1 - p_{ij}^{\text{ECM}} & \text{if } w_{ij} = 0, \\ p_{ij}^{\text{ECM}}(y_i y_j)^{w_{ij}-1}(1 - y_i y_j) & \text{if } w_{ij} > 0 \end{cases} \tag{2.31}$$

---

[5] The ECM is particularly interesting from a theoretical viewpoint. In fact, it generates mixed Bose–Fermi statistics in the case of integer weights (Garlaschelli and Loffredo, 2009), whereas in the case of continuous weights it is equivalent to a Fermi system on a lattice gas (Gabrielli et al., 2019).

with $p_{ij}^{ECM} = \frac{x_i x_j y_i y_j}{1 - y_i y_j + x_i x_j y_i y_j}$ being the binary connection probability. Examples of algorithms belonging to the second group are those using the ME recipe and then iteratively adjusting the link weights (e.g., via the IPF recipe (Bacharach, 1965)) on top of some previously determined topological structure, in such a way as to satisfy the strength constraints a posteriori.

As will be shown below, the knowledge of both the degrees and the strengths allows the ECM to achieve a very good reconstruction of many different kinds of networks (Mastrandrea et al., 2014a,b). Degrees, however, are rarely accessible, whence the need to find an alternative recipe to estimate them. As shown in (Squartini et al., 2017b), the basic information encoded into the link density of the network can be successfully used to accomplish this aim. Its use leads to the definition of a two-step approach to reconstruction.[6] This approach has been tested in four different horse races (Anand et al., 2018; Mazzarisi and Lillo, 2017; Ramadiah et al., 2020b; Lebacher et al., 2019), resulting in the method consistently performing the best (or among the best).

The first step of such an ERG-based approach consists in estimating the network topology, resting upon the following three hypotheses.

I. The binary topology of the empirical network $\mathbf{W}^*$ is drawn from the ensemble induced by the *Directed Configuration Model* (DCM) (Squartini and Garlaschelli, 2011; Park and Newman, 2004b). The DCM induces a set of configurations that are maximally random except for the (ensemble) averages of the out- and in-degrees. This amounts to considering the entries $a_{ij} = \Theta[w_{ij}]$, $\forall i \neq j$ of the binary adjacency matrix as independent random variables, fully described by the link-specific probability coefficients reading

$$p_{ij}^{DCM} = \langle a_{ij} \rangle_{DCM} = \frac{x_i y_j}{1 + x_i y_j}, \; \forall i \neq j. \tag{2.32}$$

The Lagrange multipliers $\{x_i\}_{i=1}^{N}$ and $\{y_i\}_{i=1}^{N}$ can be numerically determined by solving the system of equations

$$k_i^{out}(\mathbf{A}^*) = \sum_{j \neq i} \frac{x_i y_j}{1 + x_i y_j} = \langle k_i^{out} \rangle_{DCM}, \; \forall i \tag{2.33}$$

$$k_i^{in}(\mathbf{A}^*) = \sum_{j \neq i} \frac{x_j y_i}{1 + x_j y_i} = \langle k_i^{in} \rangle_{DCM}, \; \forall i \tag{2.34}$$

where $\mathbf{A}^* = \Theta[\mathbf{W}^*]$. Degrees, however, are rarely accessible, whence the need of a second assumption.

---

[6] As opposed to the algorithms employing the IPF recipe to adjust weights, whose second step is *deterministic*, both steps of these ERG-based approaches are *probabilistic* in nature.

II. The Lagrange multipliers $\{x_i\}_{i=1}^N$ and $\{y_i\}_{i=1}^N$ controlling for the ensemble average of the degrees are assumed to be linearly correlated with accessible quantities, generally called *fitnesses* (Caldarelli et al., 2002; Iori et al., 2008). The typical approach is to use the out- and in-strengths themselves (i.e., the only available information) as fitnesses, because of the strong correlation between degrees and strengths observed in several real-world networks (Cimini et al., 2015b). Whence the position

$$x_i = \sqrt{z} \cdot s_i^{out}, \ \forall \ i \ \text{ and } \ y_i = \sqrt{z} \cdot s_i^{in}, \ \forall \ i. \tag{2.35}$$

Since we cannot make a direct use of the DCM, we can resort to the ansatz above, which assumes the network topology to be determined by intrinsic node properties. This approach, also known as the *fitness-induced Configuration Model* (fiCM), has been successfully employed to model financial networks, where fitnesses are assets and liabilities (Cimini et al., 2015b,a). For economic networks, node fitnesses have been identified, for instance, with the countries' GDP (Garlaschelli and Loffredo, 2004a; Almog et al., 2017).

III. Besides the heterogeneity induced by the fitness-induced degrees, the network is assumed to be *homogeneous*, so that its (global) link density can be estimated by sampling subsets of nodes. To this aim, the best recipe is represented by the *random-nodes sampling scheme*, any other procedure being biased toward unrealistically large or small link density values (Squartini et al., 2017b).

The assumptions just presented leave us with the task of determining only one (global) proportionality constant, which is obtained by equating the ensemble average of the total number of links to the (known or estimated) $L(\mathbf{W}^*)$ value,

$$\langle L \rangle_{\text{fiCM}} = \sum_i \sum_{j \neq i} \frac{z s_i^{out} s_j^{in}}{1 + z s_i^{out} s_j^{in}} = L(\mathbf{W}^*); \tag{2.36}$$

once z has been determined, the numerical value of the linking probabilities,

$$p_{ij}^{\text{fiCM}} = \langle a_{ij} \rangle_{\text{fiCM}} = \frac{z s_i^{out} s_j^{in}}{1 + z s_i^{out} s_j^{in}}, \ \forall \ i \neq j, \tag{2.37}$$

can be straightforwardly estimated. Notice that the fiCM can be easily generalized to address the problem of bipartite network reconstruction (see also Appendix A) (Squartini et al., 2017a).

The second step of the reconstruction procedure then concerns the estimation of link weights. To this aim, one can extend the traditional MaxEnt prescription by defining the following Bernoulli-like recipe:

$$q_{ij}^{\text{dcGM}}(w_{ij}) = \begin{cases} 0 & \text{with probability } 1 - p_{ij}^{\text{fiCM}}, \\ \frac{\hat{w}_{ij}^{\text{ME}}}{p_{ij}^{\text{fiCM}}} & \text{with probability } p_{ij}^{\text{fiCM}}. \end{cases} \quad (2.38)$$

Since the MaxEnt weight $\hat{w}_{ij}^{\text{ME}}$ is placed between nodes i and j with probability p_{ij}^{fiCM}, both the node strengths and the link density are correctly reproduced,[7] whatever the underlying topology of the network. In fact, $\langle L \rangle_{\text{fiCM}} = L(\mathbf{W}^*)$ and

$$\langle w_{ij} \rangle_{\text{dcGM}} = 0 \cdot (1 - p_{ij}^{\text{fiCM}}) + \frac{\hat{w}_{ij}^{\text{ME}}}{p_{ij}^{\text{fiCM}}} \cdot p_{ij}^{\text{fiCM}} = \hat{w}_{ij}^{\text{ME}}, \ \forall \ i \neq j. \quad (2.39)$$

The reconstruction method described here is known as the *density-corrected Gravity Model* (dcGM) (Cimini et al., 2015b). In a sense, by disentangling the binary and weighted statistics of the ensemble, it constitutes a simplified version of the *Directed Enhanced Configuration Model* (DECM) yet retains the same accuracy in reconstructing real-world networks. Note that the dcGM approach can be "enriched" with an *exponential* weight-specific distribution that overcomes the limitations affecting the simpler Bernoulli-like recipe defined in Eq. (2.38) (Parisi et al., 2020).

An alternative, efficient approach to the dcGM consists in using degrees estimated through the first step together with empirical strengths to inform an ECM (Cimini et al., 2015a).

Testing Reconstruction at the Macroscale

This second part of the section is devoted to answering the following question: *Which of the aforementioned network patterns can be reconstructed and by which model?* As will be shown, specifying only *local* information (i.e. the information encoded into the degree and strength sequences) is often enough to achieve a satisfactory reconstruction of the network under consideration. In what follows we will focus on economic systems. In particular, we will cite results concerning the WTW, which is the network of trade exchanges between world countries. The WTW data is publicly available, allowing reconstruction models to be testable and comparable.

Assortativity and Hierarchy. Let us start by inspecting if and how assortativity can be reconstructed. Figure 1 shows the empirical ANND of the WTW for the year 2002 (Squartini et al., 2011a): The network is disassortative (i.e. countries with large degree preferentially connect with countries with low degree and vice versa); from a macroeconomic point of view, this reflects the evidence that

---

[7] In order for node strengths to be correctly reproduced, the probability distribution defined by Eq. (2.38) must include self-loops; in Squartini et al., 2017b, the authors have proposed a slightly modified version of this recipe to deal with the more realistic case in which self-loops are absent.

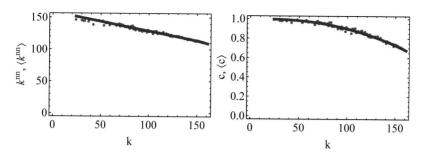

Figure 1 Comparison between the observed (red points) and the reconstructed (blue solid curve) values of the average nearest neighbors degree (left panel) and of the clustering coefficient (right panel) plotted as functions of the degree, for the 2002 snapshot of the binary undirected WTW. Because the reconstruction is done through the CM, we can conclude that the information encoded into the degree sequence is enough to reproduce the disassortative and hierarchical character of the WTW network. Source: Squartini et al., 2011a.

the partners of richer countries are (preferentially) poorer countries and vice versa. Figure 1 also shows that constraining the degrees allows degree–degree correlations to be reproduced quite well: in other words, the CM allows one to achieve an accurate reconstruction of the second-order properties of the WTW. A similar conclusion can be drawn about third-order properties such as the clustering coefficient: the CM correctly reconstructs the WTW as a hierarchical network.

Figure 2 shows the performance of the CM in reproducing the ANND and the clustering coefficient for a wide variety of networks (technological, neural, cellular, and financial ones): While the former seems to be reproduced overall quite satisfactorily, the latter is not; the best agreement is observed for a financial system, that is, e-MID (electronic Market for Interbank Deposits), the Italian unsecured interbank network (Cimini et al., 2015b). The comparison between the CM and the ER models is also shown: as anticipated, the ER model performs quite poorly in reproducing the empirical patterns considered here. This is readily seen by calculating $\langle k_i^{nn} \rangle_{\text{ER}} = p(N - 1)$, $\forall i$ and $\langle c_i \rangle_{\text{ER}} = p$, $\forall i$, a result confirming that the ER model is not able to account for the heterogeneity of nodes characterizing any real-world network.

It is interesting to notice how the weighted counterparts of the ANND and the clustering coefficient are, instead, badly reproduced by the WCM. As Fig. 3 seems to suggest, the information encoded into the strength sequence is not enough to satisfactorily reconstruct the weighted WTW structure. The reason lies in the poor performance of the WCM in reproducing the purely topological structure of the WTW: in fact, it predicts a very dense network (i.e. $\langle k_i \rangle_{\text{WCM}} \simeq$

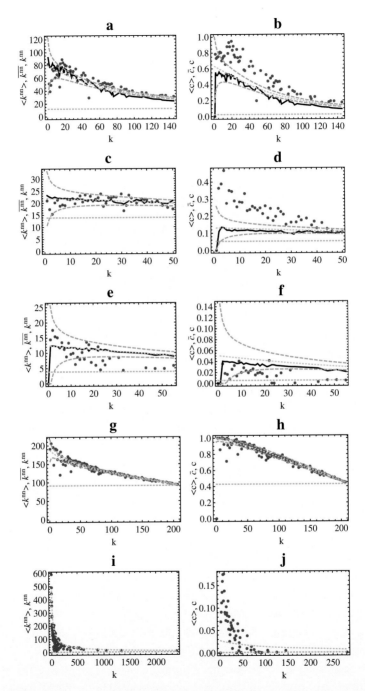

Figure 2 Average nearest neighbors degree (left panel) and clustering coefficient (right panel) of various real-world networks: the largest US airports one a), b); the C. Elegans synaptic one c), d); the E. Pylori protein-protein interaction one e), f); eMID g), h); Internet at the AS level i), j). Source: Squartini and Garlaschelli, 2011.

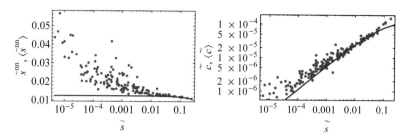

Figure 3 Comparison between the empirical (red points) and expected (blue solid curve) values of the average nearest neighbors strength (left panel) and of the weighted clustering coefficient (right panel) plotted as functions of the strength, for the 2002 snapshot of the weighted, undirected WTW (red points). The expectations are computed under the WCM: while the latter fails in reproducing the trend of the ANNS, it seems to capture the rising trend of the WCC. Source: Squartini et al., 2011b.

($N - 1$), $\forall$ i), a result inducing a flat trend of higher-order properties like the ANNS. In this respect, the ME model and the WCM perform similarly: $\langle s_i^{nn} \rangle_{ME} \simeq \langle s_i^{nn} \rangle_{WCM} \simeq \frac{2W}{N-1}$, $\forall$ i.

As mentioned in the previous section, the solution to this problem lies in constraining some kind of topological information beside the (weighted) one represented by strengths. In the ideal case, both degrees and strengths are available: constraining them simultaneously leads to the definition of the ECM (Mastrandrea et al., 2014a; Gabrielli et al., 2019), whose good performance in reproducing a wide range of real-world networks is shown in Fig. 4.

Nestedness. The upper panel of fig. 5 shows the nestedness (measured by NODF) of the bipartite WTW in the year 2000; the lower panel, instead, shows the performance of the *Bipartite Configuration Model* (BiCM – see also Appendix A) in reproducing it [Saracco et al., 2015]). The empirical and the expected values of the NODF are compared via a z-score, defined as

$$z_{NODF} = \frac{NODF - \langle NODF \rangle_{BiCM}}{\sigma_{NODF}}, \tag{2.40}$$

quantifying the difference between them in units of standard deviation.[8] As can be appreciated, $-1 < z_{NODF} < 1$, even if values $z_{NODF} \simeq \pm 2$ are, sometimes,

---

[8] An empirical value X^* corresponding to a largely positive (negative) value of z_{X^*} is assumed to indicate that the quantity X is over- or underrepresented in the data, hence not explained by the model itself. More precisely, for a quantity that is normally distributed under a given model, values falling outside the intervals $z_X = \pm 1$, $z_X = \pm 2$, $z_X = \pm 3$ occur with probabilities of 32%, 5%, 1%, respectively.

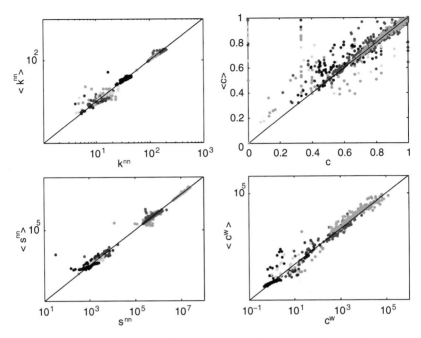

Figure 4 Enhanced network reconstruction from strengths and degrees (ECM). Each panel shows the comparison between the reconstructed (*y*-axis) and empirical (*x*-axis) values of a node-specific network property, for several real-world networks. Top left: ANND; top right: clustering coefficient; bottom left: ANNS; bottom right: WCC. Source: Mastrandrea et al., 2014a.

reached. Similar results are found in Payrató-Borràs et al., 2019 where ecological networks are considered. These results seem to indicate that the information encoded into the degree sequence is indeed enough to explain the empirical nestedness observed in economic and ecological systems – see, however, Bruno et al., 2020.

Reciprocity. Contrary to what happens for other binary properties, the DCM often fails in reproducing the empirical values of reciprocity. An example is provided by the Dutch Interbank Network (DIN), whose observed and expected reciprocities are compared in Squartini et al., 2013a. While during the first seven years covered by the dataset, the reciprocity structure of the network (inspected via its *dyadic* structure – see also Section 3) is still consistent with the DCM prediction, the remaining three years are characterized by an increasing difference between the values r and $\langle r \rangle_{\mathrm{DCM}}$.

Although this signals that the structure of the DIN cannot be fully explained by the degree heterogeneity, it also highlights the versatility of the models defined within the ERG framework: in case the chosen amount of information

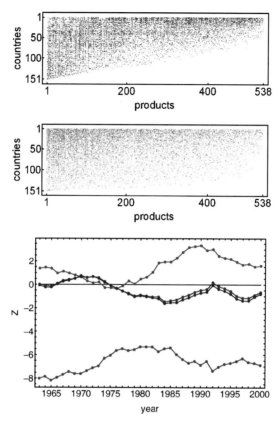

Figure 5 Upper panel: the binary, undirected, bipartite representation of the WTW in the year 2000. Upon reordering the rows and columns according to the fitness-complexity algorithm introduced in Tacchella et al., 2012, a triangular pattern clearly emerges. Middle panel: matrix drawn from the ensemble induced by the BiCM for the same year and ordered according to the same criterion. Lower panel: evolution of the z-score for the assortativity index (brown), the NODF (blue), and its "reduced" versions along rows (magenta) and columns (purple). Source: Saracco et al., 2015.

is not capable of reproducing the observations, it can still be used to define a *null model*, that is, a benchmark against which to compare the empirical patterns (Cimini et al., 2019). Alternatively, a more refined reconstruction model can be defined; in this case, explicitly constraining reciprocity together with degrees defines the *Reciprocal Configuration Model* (RCM) (Squartini and Garlaschelli, 2011; Garlaschelli and Loffredo, 2006).

Quantifying Systemic Risk

As an additional test of the goodness of the discussed reconstruction methods, let us consider the problem of quantifying systemic risk. This problem has become extremely relevant since the aftermath of the financial crisis. Systemic risk is rooted in evidence that the complex patterns of interconnections between financial institutions have the potential to make the system *as a whole* extremely fragile, as these connections constitute the channels through which financial distress can spread (Gai and Kapadia, 2010; Haldane and May, 2011; Acemoglu et al., 2015; Battiston et al., 2016; Bardoscia et al., 2017). As a consequence, both researchers and regulators have paid increasing attention to inferring the structural features of financial systems, with the aim of properly estimating the *systemicness* of an institution (Squartini et al., 2018): intuitively, a systemically important institution adversely affects a large number of other institutions in the case of default (Komatsu and Namatame, 2012). In what follows, we will provide two illustrative examples of systemic risk estimators.

Systemic Risk for Monopartite Networks. In the case of monopartite interbank networks, stress tests typically have been performed using the propagation of a shock as a consequence of the default of an institution. This is simulated by (1) deleting the defaulted institution and its connections from the network, (2) checking the impact of such an event on the other nodes, and (3) repeating the deletion step if other defaults have happened as a consequence of the previous step (Furfine, 2003).

It is, however, of greater interest to check the level of *distress* of an institution (i.e. its "closeness" to default). A compact measure in this direction is provided by the *DebtRank* (DR) indicator (Battiston et al., 2012). To obtain this indicator, the starting point is the balance sheet equation governing the financial situation of a bank i, namely $E(i) = a(i) - l(i)$, where $E(i)$ is the value of i's equity, $a(i)$ the value of its assets, and $l(i)$ the value of its liabilities. The DR of bank i, denoted as $h(i)$, is equal to $h(i) = 0$ if the bank is "healthy" (i.e. its equity is positive and has not suffered any losses); if $h(i) = 1$, bank i is defaulted (i.e. its equity is zero); the intermediate values $0 < h(i) < 1$ correspond to different levels of distress (banks have not defaulted yet but are "closer" to default as a consequence of a propagating shock).

Given the (known or reconstructed) weighted, directed adjacency matrix $\mathbf{W}$ of the network, let us call $\mathbf{E}_0$ the vector of banks equities at time $t = 0$, $\mathbf{E}_1$ the vector of banks equities at time $t = 1$, and τ the total amount of time during which the system dynamics are observed. The algorithm to calculate the DR index works as follows (Bardoscia et al., 2015):

- the equity of all banks is assumed to be affected by an external shock: as a consequence, $E_1(i) < E_0(i)$, $\forall\, i$;

- the *relative equity loss* of each bank is $h_1(i) = \frac{E_0(i)-E_1(i)}{E_0(i)} > 0$, which measures its level of distress (even if the bank has not defaulted, it has come "closer" to default as a consequence of the equity reduction);
- a distressed bank j is less likely to meet its obligations; thus, the distress of bank j becomes a distress for each bank i that lent money to j (i.e. for which $w_{ij} > 0$).
- the overall distress bank i receives at a generic time t can be calculated as

$$\Delta_t(i) = \sum_{j \neq i} \Lambda_{ij} \cdot (h_t(j) - h_{t-1}(j)), \qquad (2.41)$$

where $\Lambda_{ij} = \frac{w_{ij}}{E_0(i)}$, $\forall\, i \neq j$ is the so-called *leverage matrix* and $h_t(i) = \frac{E_{t-1}(i)-E_t(i)}{E_{t-1}(i)}$. As a consequence, the state of bank i is updated according to the rule

$$h_{t+1}(i) = \min\{1, h_t(i) + \Delta_t(i)\}; \qquad (2.42)$$

when $h(i) \geq 1$ bank i is defaulted (and remains in the "default" state at all subsequent time steps).

The DR algorithm outputs the list $h_t(i)$, $\forall\, i$ for each time step $t = 1 \ldots \tau$; in addition, it also outputs a global index (i.e. the "group" DebtRank), which is defined as the weighted average of the relative equity losses,

$$DR_t = \sum_i e(i) \cdot (h_t(i) - h_1(i)), \; \forall\, t, \qquad (2.43)$$

with $e(i) = \frac{E_0(i)}{\sum_j E_0(j)}$. In other words, $h(i)$ measures the economic value of node i that is potentially lost because of distress: notice that this is done in a recursive fashion to properly account for reverberation effects.

Figure 6 shows the performance of the dcGM in reproducing the DR index on a snapshot of the WTW and e-MID (beside other two important macroscale properties): the agreement between the dcGM-induced trend and the observed one is remarkable. As can be seen from Figure 7, this happens because of the high correlation between node degrees and strengths (used as fitnesses in the fiCM). Further analyses show that the dcGM effectively estimates the DR even when the available information is minimal (i.e. a small percentage of nodes is used to estimate the overall density of the network) (Musmeci et al., 2013; Squartini et al., 2017b).

Systemic Risk for Bipartite Networks. When considering bipartite networks as those of portfolio holdings by financial institutions, the focus is on the risk related to the sale of illiquid assets and the subsequent losses during fire sales (Shleifer and Vishny, 2011; Caccioli et al., 2014; Cont and Wagalath, 2016;

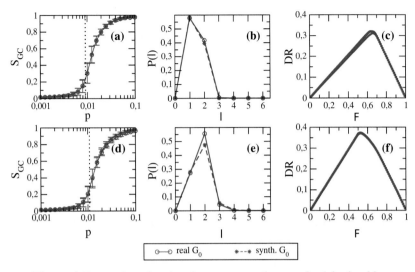

Figure 6 Properties of real and reconstructed networks (obtained by implementing the dcGM). Left plots (a,d): evolution of the size of the giant component as a function of the occupation probability. Central plots (b,e): empirical probability distribution of the directed shortest path length. Right plots (c,f): dependence of the DR on the initial distress Φ. Top panels (a,b,c) refer to WTW, bottom panels (d,e,f) to e-MID. Source: Cimini et al., 2015b.

Gualdi et al., 2016). The estimation of this kind of risk can be done by adopting the *systemicness* index S_i as a measure of the impact of a financial institute i on the whole system (Greenwood et al., 2015):

$$S_i(\mathbf{W}) = \frac{\Gamma_i V_i}{E} B_i r_i, \ \forall \ i. \tag{2.44}$$

The systemicness index is a function of $\Gamma_i = \sum_j \sum_\alpha l_\alpha (w_{i\alpha} w_{j\alpha})$, $\forall \ i$ that quantifies the overlap of portfolio i with other portfolios, the illiquidity parameter l_α of asset α, the leverage B_i and the portfolio return r_i, the total equity of the system E, and the portfolio value (i.e. the strength) of i, $V_i = \sum_\alpha w_{i\alpha}$. In order to simplify the estimation of the systemicness index, we can assume the presence of homogeneous shocks as well as identical illiquidity parameters for all assets. Upon doing so, the ratio between the expected and the observed values of systemicness becomes defined solely in terms of quantities that can be readily estimated (i.e. the link weights):

$$\frac{\langle S_i \rangle}{S_i(\mathbf{W})} = \frac{\sum_j \sum_\alpha \langle w_{i\alpha} \rangle \langle w_{j\alpha} \rangle}{\sum_j \sum_\alpha w_{i\alpha} w_{j\alpha}}, \ \forall \ i. \tag{2.45}$$

Not surprisingly, the (bipartite versions of the) dcGM (Squartini et al., 2017a) and the MECAPM (Di Gangi et al., 2018) reconstruct the same expected value

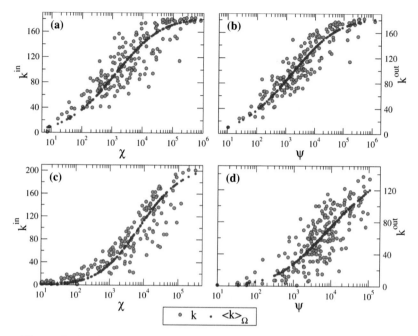

Figure 7 Qualitative assessment of the fiCM ansatz: scatter plots of node fitnesses χ, ψ (the strengths) versus real node in- and out-degrees (red circles) and their fiCM ensemble averages (blue asterisks). Upper panels (a,b) refer to WTW, lower panels (c,d) to eMID. Source: Cimini et al., 2015b.

of systemicness, which is a natural consequence of the fact that the expected values of weights under the two models coincide. This happens despite the fact that the fiCM can well reproduce degrees while MECAPM cannot – see Fig. 8.

However, results again differ when single network instances are drawn from the corresponding ensembles. Let us, in fact, replace the ensemble averages in Eq. (2.45) with the single-instance values of the weights (i.e. the weights of a particular configuration drawn from the dcGM or MECAPM ensembles),

$$\frac{\tilde{S}_i}{S_i(\mathbf{W})} = \frac{\sum_j \sum_\alpha \tilde{w}_{i\alpha} \tilde{w}_{j\alpha}}{\sum_j \sum_\alpha w_{i\alpha} w_{j\alpha}}, \ \forall \ i. \qquad (2.46)$$

As Fig. 9 shows, while the estimates provided by the two methods coincide for the largest nodes, the MECAPM tends to overestimate the systemicness of small nodes. As an additional test, let us consider the ratio of the ensemble standard deviations of the systemicness index, that is, $r_{S_i} = \frac{\sigma_{S_i}^{\text{dcGM}}}{\sigma_{S_i}^{\text{MECAPM}}}$, $\forall \ i$: it is found to be smaller than 1 for 90% of the nodes.

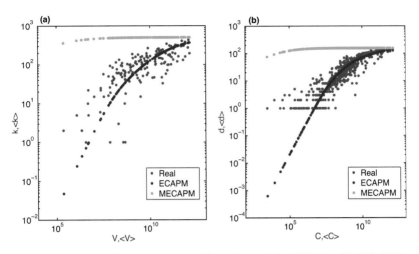

Figure 8 Performance of the (bipartite version of the) fiCM and MECAPM in reproducing the degrees of the nodes defining the Security Holding Statistics (SHS) network. Source: Squartini et al., 2017a.

The explanation for such a difference between dcGM and MECAPM resides in the different errors affecting their estimates of a generic link weight w_{ij}. In formulas, the ratio r_{ij} between the two errors is

$$r_{ij} = \frac{\sigma_{\hat{w}_{ij}^{\text{dcGM}}}}{\sigma_{\hat{w}_{ij}^{\text{MECAPM}}}} \simeq \sqrt{\frac{1}{p_{ij}^{\text{fiCM}}} - 1}, \ \forall \ i \neq j. \tag{2.47}$$

Notice the key role played by topology in lowering the uncertainty affecting the estimation of weights: requiring $r_{ij} < 1$ is, in fact, equivalent to requiring $p_{ij}^{\text{fiCM}} > 1/2$, further implying that the estimation of larger weights provided by the dcGM is less affected by uncertainty (Squartini et al., 2017b).

These results are particularly relevant for the estimation of systemic risk: Since larger weights drive larger shocks, it is desirable to employ a model providing accurate estimations for them; hence, employing the (bipartite version of the) dcGM to generate possible scenarios seems to allow obtaining configurations over which the estimation of systemic risk provides values closer to the actual ones (Squartini et al., 2017a).

3 Network Reconstruction at the Mesoscale

Reconstructing vs. Testing. When studying the mesoscale structure of a network, it is fundamental to distinguish two different, yet complementary, methodological approaches: the one seeking for the best *reconstruction* model, and the one seeking for the best *benchmark* model. In fact, testing the statistical significance of a network quantity requires the following: (i) building the benchmark by choosing some (usually local) properties of the real network to

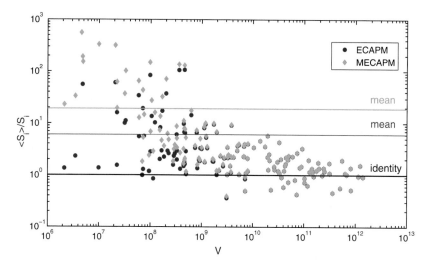

Figure 9 Reconstruction of systemicness in the SHS dataset: each point represents the relative systemicness $\frac{\tilde{S}_i}{S_i(\mathbf{W})}$ of country-sector i scattered versus its strength (here indicated with V_i) for a particular configuration drawn from the (bipartite) dcGM ensemble (blue dots) and the (bipartite) MECAPM ensemble (green dots). Averages are shown for both methods as horizontal solid lines. Source: Squartini et al., 2017a.

be preserved while randomizing everything else; (ii) comparing the empirical value of the chosen quantity with its value according to the benchmark. Finding that the considered property is *not statistically significant* ultimately means that the (local) information that has been preserved to define the benchmark is enough to explain that property: therefore, we can conclude that it is possible to *reconstruct* such a property by just knowing the aforementioned local information. On the other hand, finding *a statistically significant* discrepancy between the empirical value of the considered quantity and its randomized counterpart implies that additional information is required to explain it. As we observed in the previous section, the knowledge of *binary local information* (i.e. the degree of nodes) allows one to reconstruct several higher-order properties, while the same kind of information in the weighted case (i.e. the strength of nodes) fails. In this section we focus only on the binary part by asking *to what extent the knowledge of binary local information allows one to reconstruct the mesoscale structure of networks*.

Motifs: The Building Blocks of Networks

In the previous section, we introduced the important concept of *motif*, a specific interaction pattern involving a small number of nodes. The first work about motifs appeared in a paper by Shen-Orr et al. (2002) studying the gene regulation network of the bacteria *E. coli*. In the same year, Milo et al. (2002) presented a detailed study of the motifs in different types of real-world networks: the authors showed that these subgraphs could help in identifying *classes* of networks. Since then, an increasing body of literature has been devoted to the study of motifs in biological and neural networks (Lee et al., 2002), (Yeger-Lotem et al., 2004; Sporns and Kötter, 2004; Cloutier and Wang, 2011; Stouffer et al., 2012; Lim et al., 2013; Chen et al., 2013; Messé et al., 2018), economic systems (Ohnishi et al., 2010; Squartini and Garlaschelli, 2012; Saracco et al., 2015, 2016), and, more in general, to quantitatively characterize networks of different natures via the analysis of substructures (Watts and Strogatz, 1998; Sinatra et al., 2010; Jiang et al., 2013; Schneider et al., 2013; Stone et al., 2019).

Dyadic Motifs. In a directed binary network, the simplest motifs are represented by all possible subgraphs of dimension 2. There are four possibilities according to the presence and the direction of links between the two nodes. Given the adjacency matrix of the network, it is possible to compactly write the occurrence of each dyadic pattern as follows:

$$a_{ij}^{\rightarrow} \equiv a_{ij}(1 - a_{ji}), \tag{3.1}$$

$$a_{ij}^{\leftarrow} \equiv a_{ji}(1 - a_{ij}), \tag{3.2}$$

$$a_{ij}^{\leftrightarrow} \equiv a_{ij}a_{ji}, \tag{3.3}$$

$$a_{ij}^{\nleftrightarrow} \equiv (1 - a_{ij})(1 - a_{ji}). \tag{3.4}$$

Dyadic motifs thus refine the concept of reciprocity, a measure that simply counts the number of bilateral links with respect to the total number of links. And, despite their simplicity, these patterns can well characterize strongly symmetric networks, such as the WTW, as well as asymmetric networks, such as the investment networks of countries (Dueñas et al., 2017).

Triadic Motifs. Evidently, motifs with more than two nodes are of higher interest, but are also more complex to count, as the increasing number of nodes implies an exponential increase of the possible patterns according to the presence/absence of links and their directionality. Indeed, most of the studies in the literature focus on motifs of size 3 or 4. Despite their small size, triadic motifs offer interesting insights on network organization. Indeed, they can be considered as the natural extension of the directed clustering coefficient (number of closed triangles over all possible triplets of nodes) and represent a first step toward the deepest exploration of the network organization in communities

Motif m	N_m: 1st definition	N_m: 2nd definition
1	$\sum_{i\neq j\neq k}(1-a_{ij})a_{ji}a_{jk}(1-a_{kj})(1-a_{ik})(1-a_{ki})$	$\sum_{i\neq j\neq k}\overleftarrow{a_{ij}}\,\overrightarrow{a_{jk}}\,\overleftrightarrow{a_{ik}}$
2	$\sum_{i\neq j\neq k}a_{ij}(1-a_{ji})a_{jk}(1-a_{kj})(1-a_{ik})(1-a_{ki})$	$\sum_{i\neq j\neq k}\overrightarrow{a_{ij}}\,\overrightarrow{a_{jk}}\,\overleftrightarrow{a_{ik}}$
3	$\sum_{i\neq j\neq k}a_{ij}a_{ji}a_{jk}(1-a_{kj})(1-a_{ik})(1-a_{ki})$	$\sum_{i\neq j\neq k}\overleftrightarrow{a_{ij}}\,\overrightarrow{a_{jk}}\,\overleftrightarrow{a_{ik}}$
4	$\sum_{i\neq j\neq k}(1-a_{ij})(1-a_{ji})a_{jk}(1-a_{kj})a_{ik}(1-a_{ki})$	$\sum_{i\neq j\neq k}\overleftrightarrow{a_{ij}}\,\overrightarrow{a_{jk}}\,\overleftrightarrow{a_{ik}}$
5	$\sum_{i\neq j\neq k}(1-a_{ij})a_{ji}a_{jk}(1-a_{kj})a_{ik}(1-a_{ki})$	$\sum_{i\neq j\neq k}\overleftarrow{a_{ij}}\,\overrightarrow{a_{jk}}\,\overrightarrow{a_{ik}}$
6	$\sum_{i\neq j\neq k}a_{ij}a_{ji}a_{jk}(1-a_{kj})a_{ik}(1-a_{ki})$	$\sum_{i\neq j\neq k}\overleftrightarrow{a_{ij}}\,\overrightarrow{a_{jk}}\,\overrightarrow{a_{ik}}$
7	$\sum_{i\neq j\neq k}a_{ij}a_{ji}(1-a_{jk})a_{kj}(1-a_{ik})(1-a_{ki})$	$\sum_{i\neq j\neq k}\overleftrightarrow{a_{ij}}\,\overleftarrow{a_{jk}}\,\overleftrightarrow{a_{ik}}$
8	$\sum_{i\neq j\neq k}a_{ij}a_{ji}a_{jk}a_{kj}(1-a_{ik})(1-a_{ki})$	$\sum_{i\neq j\neq k}\overleftrightarrow{a_{ij}}\,\overleftrightarrow{a_{jk}}\,\overleftrightarrow{a_{ik}}$
9	$\sum_{i\neq j\neq k}(1-a_{ij})a_{ji}(1-a_{jk})a_{kj}a_{ik}(1-a_{ki})$	$\sum_{i\neq j\neq k}\overleftarrow{a_{ij}}\,\overleftarrow{a_{jk}}\,\overrightarrow{a_{ik}}$
10	$\sum_{i\neq j\neq k}(1-a_{ij})a_{ji}a_{jk}a_{kj}a_{ik}(1-a_{ki})$	$\sum_{i\neq j\neq k}\overleftarrow{a_{ij}}\,\overleftrightarrow{a_{jk}}\,\overrightarrow{a_{ik}}$
11	$\sum_{i\neq j\neq k}a_{ij}(1-a_{ji})a_{jk}a_{kj}a_{ik}(1-a_{ki})$	$\sum_{i\neq j\neq k}\overrightarrow{a_{ij}}\,\overleftrightarrow{a_{jk}}\,\overrightarrow{a_{ik}}$
12	$\sum_{i\neq j\neq k}a_{ij}\,a_{ji}a_{jk}a_{kj}a_{ik}(1-a_{ki})$	$\sum_{i\neq j\neq k}\overleftrightarrow{a_{ij}}\,\overleftrightarrow{a_{jk}}\,\overrightarrow{a_{ik}}$
13	$\sum_{i\neq j\neq k}a_{ij}\,a_{ji}a_{jk}a_{kj}a_{ik}\,a_{ki}$	$\sum_{i\neq j\neq k}\overleftrightarrow{a_{ij}}\,\overleftrightarrow{a_{jk}}\,\overleftrightarrow{a_{ik}}$

Figure 10 Top: graphical representation of the triadic, binary, directed motifs. Bottom: classification and definition of triadic motifs. Source: Squartini and Garlaschelli, 2012.

(Kashtan and Alon, 2005). The table depicted in Fig. 10 shows how to compute the occurrence of the 13 triadic motifs with the classical and the compact representation given by Eqs. (3.1)-(3.4).

By their nature, real networks do not necessarily contain all the triadic motifs described in the table. Indeed, Milo et al. (2002) defined the motifs as "simple building blocks" of networks, closely related to their specific functioning: the ability to process information.

Motifs in Biological Networks. Milo et al. (2002) identified classes of real networks by looking at the occurrence of motifs of size 3 and 4 with respect to properly defined null models. In particular, they tested the significance of such occurrences by employing null models preserving (i) the out- and in-degrees of nodes and (ii) the occurrence of all subgraphs of size $n-1$ when testing the significance of subgraphs of size n. This second requirement amounts to preserving the number of diadic motifs when testing the significance of triadic motifs, and the occurrence of all the 13 triadic motifs when testing the significance of motifs of size 4. This choice allows one to filter out the effect of possible significant occurrences of smaller-order motifs. The authors found that different kinds of networks are characterized by different patterns. They considered two transcriptional regulation gene networks (nodes are genes, links are directed from a gene that encodes a transcriptional factor to a gene regulated by it), finding only two significant motifs over the 13 triadic motifs and the 199 4-size motifs (directed networks): the feed-forward loop and the bi-fan. The outcome for food webs (nodes are species, and links are directed from predators to preys) of seven different ecosystems revealed the different nature of these networks, with the feed-forward loop always underrepresented, with the three chain and the bi-parallel motifs being significantly shared among them. The neuronal network (nodes are neurons connecting through synapsis) of the *C. elegans* nematode shared two motifs with the gene networks (the feed-forward loop and the bi-fan) and one with the food webs (the bi-parallel one). These results stress the great variability of patterns observable in real networks and their fundamental role in shaping their structural organization and functioning.

Another relevant work in this direction was developed in the field of neuroscience. Sporns and Kötter (2004) studied the structural motifs in brain networks (nodes are brain regions, while directed links represent anatomical connections) of macaques, cats, and *C. elegans*. They assessed the occurrence of motifs against a null model preserving the degree distribution[9] (Maslov and Sneppen, 2002). Such a null model washes away the effect of high local clustering on motifs' occurrence. For large-scale cortical networks (macaque, cat), the authors found only one statistically significant 3-node subgraph (motif 9 in Fig. 11) and its expanded versions for 4-node patterns. These motifs are relevant for explaining the cortical functional organization, as they combine the two main principles of *segregation* and *integration*: indeed, they contain reciprocal chains with unconnected end nodes. The result is different for the invertebrate *C. elegans*: other motifs appear statistically significant instead of motif 9. This outcome allows classifying the large-scale cortical networks in a separate family of brain networks with respect to the neuronal network of the nematode. In reconstruction terms, these findings imply that local node properties are not able to explain the principles driving the functional organization

---

[9] The Maslov and Sneppen (2002) randomization procedure is based on link swaps, which occur only if there are nonzero entries closer to the main diagonal in the resulting adjacency matrix: this approach produces network structures that are similar to rings or lattices.

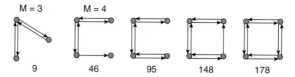

Figure 11 Significant motifs in the structural brain network. Numbers indicate the type of motif. Source: Sporns and Kötter, 2004.

of brain networks also in the simplest known neuronal system (i.e. *C. elegans*): these networks cannot be reconstructed using local information only.

Motifs in Economic and Financial Networks. More recently, Squartini and Garlaschelli (2012) and Squartini et al. (2013a) tackled the problem of motifs reconstruction in economic networks using the ERG framework. They used the *z-score* $z_X = \frac{X - \langle X \rangle}{\sigma[X]}$, where X is the occurrence of (dyadic and triadic) motifs, while $\langle X \rangle$ and $\sigma[X]$ are its expected value and standard deviation over the ensembles induced by both the *Directed Configuration Model* (DCM) and the *Reciprocal Configuration Model* (RCM).

The authors started with the WTW, an interesting case study being the result of a self-organization process driven by the global economy. As Fig. 12 shows, the number of triadic motifs cannot be reproduced by simply knowing local information (i.e. under the DCM). Indeed, independently of the year under study, the number of all motifs, except two, is underestimated in the DCM reconstructed networks. It is not surprising that the two overestimated patterns are those characterized by reciprocated links only (motifs 8 and 13 in Fig. 10). In the attempt to find the minimum amount of information able to control for the triadic properties, the authors then considered the RCM: remarkably, the dyadic information seems enough to reproduce all motifs except one (which in this case appears slightly overestimated).

The same authors found a different result when considering a financial system, namely the Dutch Interbank Network (DIN) (nodes are banks and links represent loans among them). As Fig. 13 shows, a first result concerns the existence of four different temporal profiles describing the triadic motifs' occurrence and identifying important periods related to the 2008 global financial crisis. Looking at the temporal evolution of z-scores, clear early-warning signals of the topological collapse can be appreciated both under the DCM and the RCM. An interesting interpretation for the temporal evolution of motif 9 is also provided: its overrepresentation during the period 2000–2004 signals a sort of "cyclic anomaly." This nonreciprocated loop, in fact, represents a risky configuration that may have destabilized the system well before the crisis, leading

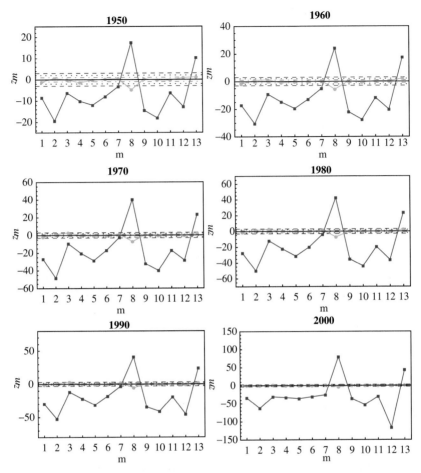

Figure 12 Evolution of the z-scores profiles for the 13 triadic motifs defined in Fig. 10, under the DCM (blue line) and the RCM (green line): while the DCM is not able to reproduce the abundance of motifs, the RCM succeeds in predicting a number of triadic motifs that is *not* significant under it. Source: Squartini and Garlaschelli, 2012.

to a lack of trust between banks during the years 2005–2007. According to this viewpoint, the crisis represented the ending point of a process that involved the entire Dutch system for years.

Bipartite Motifs. The concept of motif can be also extended to bipartite networks. As explained in the previous section, in a bipartite network there are two disjoint and independent sets of nodes, and links can exist only between and not within the two sets. It is evident that odd cycles of any length are absent, therefore neither the clustering coefficient nor the standard triadic motifs can be observed in such systems. However, in the spirit of the monopartite case, it is

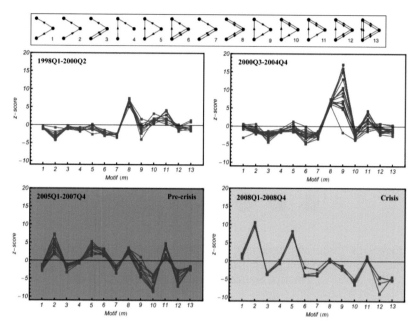

Figure 13 Evolution of the z-scores profiles for the 13 motifs defined in Fig. 10, under the DCM: notice the presence of four different temporal profiles characterizing the DIN. Source: Squartini et al., 2013a.

possible to define a class of motifs able to capture the higher-order correlations between nodes in bipartite networks.

There are examples in different fields of study. For instance, Baker et al. (2015) identified 44 motifs in a bipartite host-parasitoid food web made up of 2–6 species and uniquely identified positions. Saracco et al. (2015) considered the bipartite version of the WTW (i.e. the set of world countries and the set of products exported) and introduced the V- and Λ-motifs, respectively, counting how many pairs of countries export the same product and how many pairs of products are in the basket of the same country. In other words, the V-motif measures the correlation among producers, while the Λ-motif focuses on correlations between products (see Fig. 14, top):

$$N_V(\mathbf{B}) = \sum_i \sum_{j>i} C_{ij} = \sum_i \sum_{j>i} \sum_\alpha b_{i\alpha} b_{j\alpha} = \sum_\alpha \binom{h_\alpha}{2}, \qquad (3.5)$$

$$N_\Lambda(\mathbf{B}) = \sum_\alpha \sum_{\beta>\alpha} P_{\alpha\beta} = \sum_\alpha \sum_{\beta>\alpha} \sum_i b_{i\alpha} b_{i\beta} = \sum_i \binom{k_i}{2} \qquad (3.6)$$

where $b_{i\alpha}$ is the generic entry of the biadjacency matrix (see Appendix A). While C_{ij} counts the number of products exported by both countries i and j, $P_{\alpha\beta}$ counts the number of countries exporting both products α and β. Notice

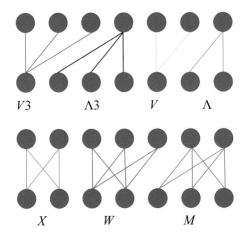

Figure 14 Undirected bipartite motifs containing up to five nodes. Top: *V*- and Λ-motifs and their generalizations to three links. Bottom: *X*, *W*, and *M* motifs. Source: Saracco et al., 2015.

that the abundance of V- and Λ-motifs can be compactly written in terms of node degrees $k_i(\mathbf{B}) = \sum_\alpha b_{i\alpha}$ (also called *diversification* and measuring the number of products exported by each country) and $h_\alpha(\mathbf{B}) = \sum_i b_{i\alpha}$ (also called *ubiquity* and measuring the number of countries exporting each product).

V- and Λ-motifs can be generalized to include more than two products/countries: for example, V3- and Λ3-motifs quantify, respectively, the number of country triplets that export the same products and how many triplets of products are in the same basket of a producer; even more generally, formulas 3.5 and 3.6 can be extended to compute the occurrence of *n*-tuples of countries and products, that is,

$$N_{V_n}(\mathbf{B}) = \sum_\alpha \binom{h_\alpha}{n} \quad \text{and} \quad N_{\Lambda_n}(\mathbf{B}) = \sum_i \binom{k_i}{n}. \tag{3.7}$$

More-complicated motifs capturing higher-order correlations among nodes can be defined by increasing the size of the subgraph. This is, for example, the case of X-, W-, and M- motifs (see Fig. 14, bottom). X-motifs measure the co-occurrence of two countries in producing the same pair of products and the co-existence of two products in the basket of the same two countries. This allows quantifying the competitiveness among countries for different market segments. M- and W- motifs allow the competitiveness between countries to be measured with respect to larger baskets of products:

$$N_X(\mathbf{B}) = \sum_{i<j} \sum_{\alpha<\beta} b_{i\alpha} b_{i\beta} b_{j\alpha} b_{j\beta} = \sum_{i<j} \binom{C_{ij}}{2} = \sum_{\alpha<\beta} \binom{P_{\alpha\beta}}{2}, \tag{3.8}$$

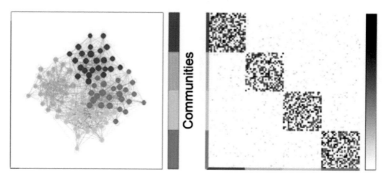

Figure 15 Examples of network community structure and its adjacency matrix representation. Source: Faskowitz et al., 2018.

$$N_M(\mathbf{B}) = \sum_{i<j} \sum_{\alpha<\beta<\gamma} b_{i\alpha} b_{i\beta} b_{i\gamma} b_{j\alpha} b_{j\beta} b_{j\gamma} = \sum_{i<j} \binom{C_{ij}}{3}, \qquad (3.9)$$

$$N_W(\mathbf{B}) = \sum_{\alpha<\beta} \sum_{i<j<k} b_{i\alpha} b_{i\beta} b_{j\alpha} b_{j\beta} b_{k\alpha} b_{k\beta} = \sum_{\alpha<\beta} \binom{P_{\alpha\beta}}{3}. \qquad (3.10)$$

Community Structure

A network is characterized by a community structure if it contains groups of nodes (the communities, or modules) clearly identifiable/recognizable for sharing common properties. The simplest possible classification looks at the number of links within and between communities. In some sense, this generalizes the concept of motifs focusing on subgraphs of higher size and looking at their internal link density rather than at specific patterns.

Let us consider a binary, undirected graph with N nodes and a subgraph C of N_C nodes. The internal density $\delta_{int}(C)$ and the external density $\delta_{ext}(C)$ of C are defined as

$$\delta_{int}(C) = \frac{\# \text{ internal links of } C}{N_C(N_C-1)/2}; \quad \delta_{ext}(C) = \frac{\# \text{ external links of } C}{N_C(N-N_C)}. \qquad (3.11)$$

A first, intuitive, requirement to identify C as a community is that $\delta_{int}(C) \gg \rho$ and $\delta_{ext}(C) \ll \rho$, where ρ is the density of the whole network (see Fig. 15 for an example illustration). Searching for the best trade-off between the two constraints is at the basis of most community detection techniques.

Investigating the community structure of a network can be relevant for a number of reasons. First of all, grouping nodes sharing some properties leads to a meta-scale (coarse-grained) representation of the network, allowing a simplified analysis (Palla et al., 2005). Moreover, the presence of communities exhibiting different properties can reveal novel features of the network with respect to its average properties, because the macroscale level does not necessarily reflect the mesoscale one. Communities could also provide insights on network function as they can act as different specialized units, as in protein-to-protein interactions (Chen and Yuan, 2006) or metabolic cycles and pathways (Guimera and Amaral, 2005). Furthermore, they can help to classify vertices according to their role within and between modules with a consequent effect on network control and stability (Csermely, 2008). They can relevantly shape diffusion patterns on networks (contagion, rumor spreading, innovation adoption). Finally, they can be used to identify nodes sharing similar interests or opinions and thus to set up efficient recommendation systems (Reddy et al., 2002).

The community detection problem still represents an open issue. Many algorithms have been introduced in the last two decades – see Fortunato (2010) and Fortunato and Hric (2016) for extensive reviews on existing approaches. It is possible to distinguish two main kinds of definitions to describe communities: the *local* and the *global* ones.

Local definitions are based on internal properties of communities such as degree, mutuality, reachability, and internal/external cohesion. Cliques (i.e, fully connected subgraphs) or clique-like modules belong to this group of definition. In the same spirit, it is possible to associate to each module a *fitness measure* to quantify how well it is defined: for example, requiring that C is a community if its internal density is larger than a fixed threshold.

Global definitions are used when the communities forming a network cannot be considered as independent entities, and a measure taking into account the whole network structure is necessary. In this case, the standard procedure prescribes to compare the empirical features of a given network partition with the value of the same features provided by a benchmark: the larger the deviation from the benchmark, the stronger the presence of a community structure. The popular *modularity* functional belongs to this class of definitions, since it compares the link density of a given partition and the expected link density of the same partition under a null model:

$$Q = \frac{1}{2L} \sum_i \sum_{j \neq i} (a_{ij} - p_{ij})\delta_{g_i g_j}; \qquad (3.12)$$

in the definition above, L is the total number of links, p_{ij} represents the probability that nodes i and j are connected under the chosen null model, g_i represents the group to which node i belongs and analogously for g_j, and the δ_{xy} function stands for the Kronecker delta, which is equal to 1 if $x = y$ (i.e. if i and j belong to the same community), and to 0 otherwise.

When binary networks are considered, the most popular benchmark is the Chung-Lu (CL) model (Chung and Lu, 2002), which preserves the degrees of nodes. Equation (3.12), thus, becomes

$$Q = \frac{1}{2L} \sum_i \sum_{j \neq i} \left(a_{ij} - \frac{k_i k_j}{2L} \right) \delta_{g_i g_j}. \tag{3.13}$$

An alternative choice of benchmark is the *Configuration Model* (CM), which, as we have seen in the previous section, is the model preserving the degrees in a proper and unbiased way. Notice that this kind of local information allows a wide set of other higher-order properties to be reconstructed (Squartini et al., 2011a; Mastrandrea et al., 2014a; Squartini et al., 2015), hence representing a nontrivial benchmark against which to compare a network partition. Indeed, the modularity maximization identifies the best partition as the one that maximally deviates from the benchmark, that is, the modular structure that is least likely to be reconstructed by knowing only local information: in other terms, a modularity value close to 0 indicates the presence of a community structure that can be inferred by just knowing the degrees of nodes.[10]

The Stochastic Block Model

From a network reconstruction perspective, the problem of inferring communities can be restated as the problem of finding the model *best fitting a given community structure*. This approach requires a certain amount of information concerning the partition of nodes into modules to be explicitly included into the model *ab initio*. Luckily, this can be done within the ERG framework (Fronczak et al., 2013). To this aim, one of the most used benchmarks is the so-called *Stochastic Block Model* (SBM) (Fienberg and Wasserman, 1981; Holland et al., 1983; Snijders and Nowicki, 1997), prescribing that the probability of connection between nodes i and j depends only on the modules (or groups) they belong to:

$$p_{ij} \equiv p_{g_i g_j}. \tag{3.14}$$

The Hamiltonian to consider, thus, becomes

$$H(\mathbf{A}) = \sum_i \sum_{j > i} \theta_{g_i g_j} a_{ij} = \sum_r \sum_{s \geq r} \theta_{rs} L_{rs}(\mathbf{A}), \tag{3.15}$$

---

[10] Notice that Q is always smaller than 1 but can admit negative values – for example, for the trivial partition where each node is considered as a module on its own. It is also worth noticing that the maximum modularity grows with the network size: therefore, it cannot be used to compare partitions of networks having a different number of nodes.

where L_{rs} represents the number of links between groups r and s (or within the same group, in case $r = s$). The partition function is thus

$$Z(\vec{\theta}) = \prod_r \left(1 + e^{\theta_{rr}}\right)^{\binom{N_r}{2}} \prod_t \prod_{s>t} \left(1 + e^{\theta_{ts}}\right)^{N_t N_s}, \qquad (3.16)$$

where N_r is the number of nodes in group r. The likelihood maximization prescription allows calculating the parameters as follows:

$$p_{rr} = \frac{L_{rr}(\mathbf{A}^*)}{\binom{N_r}{2}}, \ \forall\, r \qquad (3.17)$$

$$p_{ts} = \frac{L_{ts}(\mathbf{A}^*)}{N_t N_s}, \ \forall\, t < s. \qquad (3.18)$$

In other words, solving the SBM amounts to solving the ER model *within* each block and *between* blocks: in fact, Eq. (3.18) is nothing else than the prescription defining the bipartite ER model. Despite its simplicity, this model allows reconstructing different network structures (e.g. networks with disconnected components, core-periphery, hierarchical, or traditional modular structures).

The SBM, however, suffers from a limitation that is similar to the one affecting the ER model: in fact, the degree distribution predicted by the SBM is homogeneous and deviates from what observed in real-world networks. As nodes' heterogeneity is fundamental for correctly understanding important network properties as their resilience to external shocks, the threshold of the percolation transition, or the outcome of an epidemic spreading, Karrer and Newman (2011) proposed to incorporate in the model the information about the node degrees *beside* that concerning their group membership. In this way, they introduced a variation of the SBM, namely the *degree-corrected Stochastic Block Model* (dcSBM). In a nutshell, while the SBM is defined by connection probabilities

$$p_{g_i g_i} = \frac{e^{-\theta_{g_i g_i}}}{1 + e^{-\theta_{g_i g_i}}} \equiv \frac{\chi_{g_i g_i}}{1 + \chi_{g_i g_i}}, \ \forall\, i < j \qquad (3.19)$$

(i.e. encoding information only about the group membership of each node), the dcSBM adds the degree information by considering probability coefficients reading

$$p_{ij} = \frac{e^{-(\theta_i + \theta_j + \theta_{g_i g_i})}}{1 + e^{-(\theta_i + \theta_j + \theta_{g_i g_i})}} \equiv \frac{x_i x_j \chi_{g_i g_i}}{1 + x_i x_j \chi_{g_i g_i}}, \ \forall\, i < j, \qquad (3.20)$$

according to which the probability that nodes i and j are linked depends *both* on their group membership *and* on their degree. In fact, the Hamiltonian can now be written as:

$$H(\mathbf{A}) = \sum_i \theta_i k_i + \sum_i \sum_{j>i} \theta_{g_i g_j} a_{ij}, \qquad (3.21)$$

and the likelihood maximization prescribes to solve the following system of equations to determine the unknown parameters

$$k_i(\mathbf{A}^*) = \sum_{j \neq i} \delta_{g_i r} \delta_{g_j s} \frac{x_i x_j \chi_{rs}}{1 + x_i x_j \chi_{rs}} = \langle k_i \rangle_{\text{dcSBM}}, \ \forall \ i \qquad (3.22)$$

$$L_{rs}(\mathbf{A}^*) = \sum_i \sum_{j>i} \delta_{g_i r} \delta_{g_j s} \frac{x_i x_j \chi_{rs}}{1 + x_i x_j \chi_{rs}} = \langle L_{rs} \rangle_{\text{dcSBM}}, \ \forall \ r \leq s. \quad (3.23)$$

Notice that posing

$$\theta_i + \theta_j + \theta_{g_i g_j} = \left(\theta_i + \frac{\theta_{g_i g_j}}{2}\right) + \left(\theta_j + \frac{\theta_{g_i g_j}}{2}\right) \equiv \theta_i^{g_i g_j} + \theta_j^{g_i g_j} \qquad (3.24)$$

leads to a degree-informed model where *block-specific degrees* are constrained. In fact, the above position induces the following probability coefficients:

$$p_{ij} = \frac{e^{-\left(\theta_i^{g_i g_j} + \theta_j^{g_i g_j}\right)}}{1 + e^{-\left(\theta_i^{g_i g_j} + \theta_j^{g_i g_j}\right)}} \equiv \frac{x_i^{g_i g_j} x_j^{g_i g_j}}{1 + x_i^{g_i g_j} x_j^{g_i g_j}}, \ \forall \ i < j, \qquad (3.25)$$

defining what is known as the *Block Configuration Model* (BCM). The latter, in turn, defines the following recipe for likelihood maximization:

$$k_i^{rs}(\mathbf{A}^*) = \sum_{j \neq i} \delta_{g_i r} \delta_{g_j s} \frac{x_i^{rs} x_j^{rs}}{1 + x_i^{rs} x_j^{rs}} = \langle k_i^{rs} \rangle_{\text{BCM}}, \ \forall \ i, \ \forall \ r \leq s. \qquad (3.26)$$

In other words, solving the BCM amounts to solving the undirected version of the CM *within each diagonal block* and the *Bipartite Configuration Model* (BiCM – also Appendix A) *within each off-diagonal block*.

In the same context of the SBM, Peixoto (2017) introduced a *nonparametric* Bayesian method to infer the modular structure of a network without any *a priori* information. Indeed, like any other parameter of the model as the node membership, the optimal number of communities also is inferred by the data themselves. The procedure (see a schematic representation in Fig. 16) is based on the *microcanonical* formulation of the degree-corrected SBM (i.e. the degree sequence is fixed exactly and not on average).

The Core-Periphery Organization

The notion of *core-periphery*, a structure consisting of a bulk of densely connected nodes and a periphery of weakly linked nodes, has a long tradition in social studies (Laumann and Pappi, 1976; Doreian, 1985) and was first formalized by Borgatti and Everett (2000). The core-periphery organization has

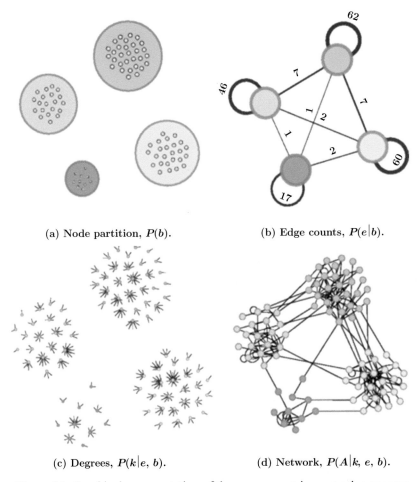

(a) Node partition, $P(b)$. (b) Edge counts, $P(e|b)$.

(c) Degrees, $P(k|e, b)$. (d) Network, $P(A|k, e, b)$.

Figure 16 Graphical representation of the nonparametric generative process for the degree-corrected SBM: (a) partition sampled; (b) link-counts between groups; (c) node degrees; (d) the network itself. Source: Peixoto, 2017.

been detected in different kinds of networks: economic (Smith and White, 1992; Veld and van Lelyveld, 2014; Fricke and Lux, 2015; Ma and Mondragón, 2015; Barucca and Lillo, 2018; Kojaku et al., 2018; van Lidth de Jeude et al., 2019b,a), social (Everett and Borgatti, 1999; Holme, 2005; Boyd et al., 2006; Della Rossa et al., 2013; Csermely et al., 2013; Zhang et al., 2015; Rombach et al., 2017; Kojaku and Masuda, 2017), biological (Yang and Leskovec, 2014; Bruckner et al., 2015), neural (Bassett et al., 2013; Tunç and Verma, 2015), and transportation networks (Lee et al., 2014; Xiang et al., 2018).

The adjacency matrix of a core-periphery network can be rearranged as a 4-block matrix (Rombach et al., 2014):

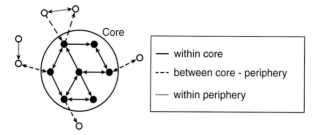

Figure 17 Three kinds of links characterize a core-periphery structure: the ones within the core, the ones between the core and the periphery, and the ones within the periphery.

$$\mathbf{A} = \begin{pmatrix} \mathbf{A}^{\bullet} & \mathbf{A}^{\top} \\ \mathbf{A}^{\perp} & \mathbf{A}^{\circ} \end{pmatrix}, \qquad (3.27)$$

where $\mathbf{A}^{\bullet}$ is associated to the core subgraph, $\mathbf{A}^{\circ}$ to the periphery subgraph, and $\mathbf{A}^{\top}$ and $\mathbf{A}^{\perp}$ contain all the connections between the two groups of nodes. It is worth noticing that while the two diagonal blocks are squared matrices, the off-diagonal ones are generally rectangular. Furthermore, generally the densities of the blocks satisfy the following chain of inequalities:

$$\rho(\mathbf{A}^{\bullet}) > \rho(\mathbf{A}^{\top}) \simeq \rho(\mathbf{A}^{\perp}) > \rho(\mathbf{A}^{\circ}), \qquad (3.28)$$

with the core being much denser than the periphery subgraph. Even if $\mathbf{A}^{\top}$ and $\mathbf{A}^{\perp}$ have a similar density, they give different information about the network structure, as the former contains all links pointing from the core to the periphery, while the latter takes into account the inverse direction. It is, thus, evident that three kinds of links are required to fully describe such a network structure: connections *within* the core, *within* the periphery, and *between* them (Fig. 17).

The issue of identifying core-periphery structures has been tackled using three different approaches: (1) minimizing the distance with respect to an ideal core-periphery structure (Borgatti and Everett, 2000; Veld and van Lelyveld, 2014); (2) defining a proper benchmark against which to detect a statistically significant topology (Holme, 2005; Della Rossa et al., 2013; Kojaku and Masuda, 2017); (3) finding the model best fitting a given mesoscale structural organization (Zhang et al., 2015; Barucca and Lillo, 2016, 2018). As for the issue of community detection, the third approach is the one that can be better framed within the network reconstruction perspective.

Borgatti and Everett (2000) first formalized the concept of core-periphery introducing the *ideal core-periphery structure* (i.e. a fully connected core and a periphery of nodes linked only with the core ones – see Fig. 18) and measuring

	1	1	1	1	0	0	0	0	0
1		1	1	0	1	1	1	0	0
1	1		1	0	0	0	1	1	0
1	1	1		1	0	0	0	0	1
1	0	0	1		0	0	0	0	0
0	1	0	0	0		0	0	0	0
0	1	0	0	0	0		0	0	0
0	1	1	0	0	0	0		0	0
0	0	1	0	0	0	0	0		0
0	0	0	1	0	0	0	0	0	

1		1	1	1	1	1	1	1	1	1
2	1	1	1	1	1	1	1	1	1	1
3	1	1	1	1	1	1	1	1	1	1
4	1	1	1	1	1	1	1	1	1	1
5	1	1	1	1	1	0	0	0	0	0
6	1	1	1	1	0	1	0	0	0	0
7	1	1	1	1	0	0	1	0	0	0
8	1	1	1	1	0	0	0	1	0	0
9	1	1	1	1	0	0	0	0	1	0
10	1	1	1	1	0	0	0	0	0	1

Figure 18 Adjacency matrices of an undirected network with a (noisy) core-periphery structure (left) and an ideal core-periphery structure (right). Source: Borgatti and Everett, 2000.

the deviation of an observed real network structure from it. In other words, the authors proposed to solve a maximization problem whose score function reads

$$\Phi = \sum_i \sum_{j \neq i} a_{ij} \Delta_{ij}, \qquad (3.29)$$

where a_{ij} is the adjacency matrix of the network, and Δ_{ij} represents the ideal core-periphery organization of a network of the same size. Since $\Delta = \delta^T \delta$, where δ is a Boolean vector whose i-th entry is equal to 1 if node i belongs to the core and 0 if it does not, maximizing Eq. (3.29) means finding a vector δ that maximizes the correlation between Δ_{ij} and a_{ij}. However, as the authors explicitly stated, a significance test for the algorithm output is completely missing.

Along the same guidelines is the work by Veld and van Lelyveld (2014). They tested the goodness of three models in recovering the core-periphery structure of the DIN as defined by some axioms: the ER, the CM, and the Barabasi–Albert (BA) model. As an accuracy index, the authors used an error score counting the number of errors (i.e. the number of links to add/delete to recover the aforementioned axiomatic model) divided by the effective number of links. They found the error score characterizing both ER and BA models was too large for the DIN structure to be compatible with them; however, the error score was not significant under the CM, a result implying that the knowledge of the out- and in-degrees allows the core-periphery mesoscale organization to be recovered to a very good extent. This confirmed earlier results by Lip (2011) that, at least for the simplest specification of the error score given in Borgatti and Everett (2000), the core-periphery partition is completely determined by the degree sequence.

Let us now discuss the algorithms belonging to the second group, namely the ones not assuming the existence of an ideal core-periphery structure but comparing the observed topology with the outcome of a properly chosen benchmark model. The first model in this sense has been proposed by Holme (2005) who introduced a generalization of the closeness centrality to be compared with a null model preserving the node degrees. The generalized closeness centrality refers to a subset U of the set of nodes V of a network

$$C_C(U) = \left[\overline{\left(\overline{d(i,j)}_{j \in V \setminus \{i\}} \right)}_{i \in U} \right]^{-1}, \qquad (3.30)$$

where $d(i,j)$ is the geodesic distance between nodes i and j. The author searched for the optimization of the score function

$$C_{cp} = \frac{C_C[V_{core}]}{C_C[V]} - \left\langle \frac{C_C[V_{core}]}{C_C[V]} \right\rangle, \qquad (3.31)$$

where the subset $U = V_{core}$ was taken to be the k-core, representing the maximal subgraph of the network having minimum degree k and maximal closeness, and the average was taken over the ensemble of networks having the same degree sequence as the observed one. Obviously, random graphs have (on average) C_{cp} equal to 0; positive/negative values of C_{cp} stand for over/underrepresentation of the core-periphery structure. The authors find that the core-periphery structure clearly characterizes groups of networks (see Table 19): geographically embedded networks often exhibit a positive C_{cp}, and this could be the effect of the optimization of temporal communications; on the other hand, social networks tend to show a slightly negative coefficient. Upon considering the specific datasets under study, the authors concluded that the existence of modules, according to some kind of specialization, could imply the absence of a clearly defined core.

Della Rossa et al. (2013) proposed an algorithm based on the standard random walk to identify the core-periphery structure of all kinds of networks, including the weighted ones. The authors associated to each network a *core-periphery profile*, namely a discrete vector $\{\alpha_1, \alpha_2 \ldots \alpha_N\}$, with N being the network size, which allows one to quantify to what extent a network is centralized and to associate to each node a measure of *coreness*. Thanks to this definition, it is possible to introduce an α-dependent degree of *peripheryness*, composed of nodes whose coreness is below a certain threshold α.

Recently, van Lidth de Jeude et al. (2019a) proposed a benchmark model for core-periphery detection inspired by the definition of *surprise*.[11] The idea is to detect *bimodular structures*, such as the core-periphery one, by comparing the probability assigned to them by the *Directed Erdös–Renyi Model* (DER) and by

---

[11] The surprise is defined as the p-value of an hypergeometric distribution (Nicolini and Bifone, 2016).

Network		N	M	C_{cp}
Geographical networks	Interstate highways	935	1315	0.231(1)
	Pipelines	2999	3079	0.180(2)
	Streets, Stockholm	3325	5100	0.255(1)
	Streets, Göteborg	1258	1516	0.040(3)
	Airport	449	2795	0.0523(3)
	Internet	1968(66)	4051(121)	0.045(2)
One-mode projections of	arXiv	48561	287570	−0.08(3)
affiliation networks	Board of directors	6193	43074	−0.037(2)
	Ajou University students	7285(128)	75898(6566)	−0.08(1)
Acquaintance networks	High School friendship	571(43)	1078(85)	0.006(7)
	Prisoners	58	83	−0.043(2)
	Social scientists	34	265(35)	−0.002(4)
Electronic communication	e-mail, Ebel *et al.*	39592	57703	−0.229(4)
	e-mail, Eckmann *et al.*	3186	31856	−0.091(2)
	Internet community, nioki.com	49801	239265	−0.014(2)
	Internet community, pussokram.com	28295	115335	−0.183(5)
Reference networks	WWW, nd.edu	325729	1090108	−0.027(3)
	HEP citations	27400	352021	−0.10(1)
Software dependencies	GNU/Linux	504	793	−0.155(1)
Food webs	Little Rock Lake	92	960	0.005(6)
	Ythan Estuary	134	593	−0.020(1)
Neural network	*C. elegans*	280	1973	0.040(6)
Biochemical networks	*Drosophila* protein	2915	4121	−0.035(2)
	S. cervisiae protein	3898	7283	−0.249(1)
	S. cervisiae genetic	1503	5043	−0.0646(7)
	Metabolic networks	427(27)	1257(88)	−0.002(6)
	Whole cellular networks	623(32)	1752(103)	−0.004(6)

Figure 19 Network size (number of nodes, N, and links, M) and core-periphery coefficient (C_{cp}) for different networks. Source: (Holme, 2005)

the SBM: finding the most statistically significant partition ultimately means finding the partition that is least likely to be explained by the DER with respect to the SBM. As shown in Fig. 20, this approach allowed identifying significant core-periphery structures in several real networks.

Similarly, Kojaku and Masuda (2017) proposed a model to identify multiple core-periphery structures by extending the approach introduced in Borgatti and Everett (2000) and using a random graph to test the significance of such a mesoscale organization. They also showed that the information encoded into the degree sequence always accounts for the organization of a network into a core and a periphery: therefore, it is not possible to use the CM to test the significance of such a mesoscale structure.

Lastly, let us discuss the approaches belonging to the third group. Zhang et al. (2015) proposed a method to identify the generative model most likely to produce a given network partition and applied it to some synthetic systems and two empirical networks: the Internet at the level of autonomous system, and the US political blogosphere (Adamic and Glance, 2005). The authors employed an expectation-maximization (EM) algorithm, assuming that the networks were generated by a SBM. Initially there are N nodes, no links, and two empty groups (the core and the periphery). Each node is randomly assigned to group 1 with

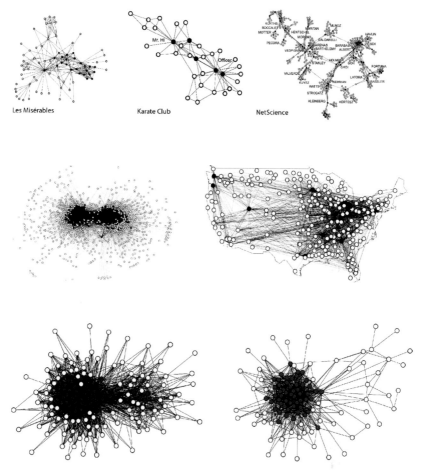

Figure 20 Core-periphery structure of several networks detected using the multivariate extension of surprise. In the top panel, "Les Miserables" (left), the Zacary Karate Club (middle), the NetSci network of collaborations (right); in the middle pane: the US political weblog (left), the US airport network (right); in the bottom panel: eMID in January 2005 (left) and November 2009. Source: van Lidth de Jeude et al., 2019a.

probability γ_1 (and to group 2 with probability $\gamma_2 = 1 - \gamma_1$). Then, each pair of nodes is connected with probability p_{rs}, where r, s indicate the groups they belong to. Given the adjacency matrix $\mathbf{A}$, the likelihood that the network is generated by the model above is given by

$$P(\mathbf{A}|p, \vec{\gamma}) = \sum_g P(\mathbf{A}|p, \vec{\gamma}, g)P(g|\vec{\gamma})$$

$$= \sum_g \left[\prod_{i<j} p_{g_i g_j}^{a_{ij}} (1 - p_{g_i g_j})^{1-a_{ij}} \prod_i \gamma_{g_i} \right], \quad (3.32)$$

where again g_i represents the group node i belongs to and $\sum_g$ is the sum over all assignments of the nodes to groups. Upon maximizing the likelihood score function $\ln P(\mathbf{A}|p, \vec{\gamma})$ it is possible to determine the values of $\vec{\gamma}$ and p_{rs}.

A few years later, Barucca and Lillo (2016, 2018) developed a SBM able to reproduce a bipartite or a core-periphery structure through a tuning parameter. They tested the e-MID structure to understand if it is better represented by a core-periphery or a bipartite structure. This work provided yet another piece of evidence of the importance of degree heterogeneity and the information it carries about the system. Indeed, the authors found that while the SBM identifies a core-periphery structure for e-MID, the dcSBM highlights a bipartite organization.

van Lidth de Jeude et al. (2019b) focused on the DIN and tested the accuracy of a plethora of methods in reproducing its core-periphery structure: in particular, they compared the block models (SBM and dcSBM) with the DCM and the RCM, finding that the DCM always performs best. This result, in line with that of Kojaku and Masuda (2018), further confirms the high informative role of node degrees in reproducing the core-periphery structure of a network.

The Bow-Tie Organization

The first formal definition of a bow-tie network decomposition was given by Yang et al. (2011), even if its concept was firstly introduced by Broder et al. (2000) for the study of the WTW. The bow-tie decomposition has, since then, been investigated both theoretically and empirically with applications in different fields: Dill et al. (2002) studied the WWW self-similarity; Arasu et al. (2002) used the bow-tie structure as a model for the large-scale structure of the WWW to test the PageRank algorithm; Hirate et al. (2008) studied the temporal evolution of bow-tie structures; Zhang et al. (2007) focused on the bow-tie structure of the Java Developer Forum to test some ranking algorithms for the expertise network; Tanaka et al. (2005) showed that metabolic networks also could be characterized by such a mesoscale structure.

The definition of a bow-tie structure is strictly related to that of node *reachability*. A node i is reachable by a node j if a path of consecutive links starting from node j and ending at node i exists. Using this concept, it is possible to fully describe a bow-tie structure as a composition of specific, non-overlapping network subgraphs (Broder et al., 2000). We report here the definition of the three most relevant subgraphs (see Fig. 21):

- the Strongly Connected Component (SCC) contains all nodes reachable by any other node in the SCC;

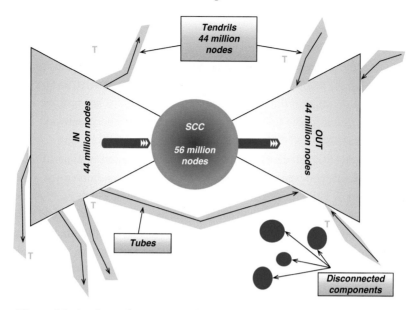

Figure 21 A schematic representation of a bow-tie structure. It contains the SCC, IN, and OUT components described in the main text. Moreover, there are (i) *tendrils* containing nodes reachable by the IN component or that can reach the OUT component without passing through the SCC one; (ii) *tubes* allowing the passage from the IN to the OUT component without passing through the SCC one. Source: Yang et al., 2011.

- the IN component (IN) contains all nodes not belonging to the SCC but from which all nodes in the SCC are reachable;
- the OUT component (OUT) contains all nodes not belonging to the SCC but reachable from any other node in the SCC.

The adjacency matrix associated with a bow-tie structure can be rearranged in order to separate the three components described above:

$$\mathbf{A} = \begin{pmatrix} \mathbf{A}^I & \mathbf{A}^> & 0 \\ 0 & \mathbf{A}^S & \mathbf{A}^\gg \\ 0 & 0 & \mathbf{A}^O \end{pmatrix}, \tag{3.33}$$

where $\mathbf{A}^I$, $\mathbf{A}^S$, and $\mathbf{A}^O$ represent respectively the IN, SCC, and OUT components, while $\mathbf{A}^>$ and $\mathbf{A}^\gg$ stand for the bipartite networks of their interactions.

The same considerations guiding the analysis of the core-periphery structure can be considered as valid also when analyzing the bow-tie structure. Let us now revise the approaches that have been proposed to study it.

Zhao et al. (2007) explored the gene-based metabolic network of 75 organisms, comparing their topological properties with a randomized counterpart preserving the degree of nodes and the number of bidirectional links. They found that the global bow-tie structure is still present in the randomized model, but with some differences: (i) the size of the SCC is smaller in the reconstructed network; (ii) the number of 2-cores is overestimated by the model, and (iii) no 3-cores are detected in the random network. These results are in favor of a significant cliquish bow-tie topology characterizing metabolic networks that cannot be reconstructed by enforcing local constraints only.

Vitali et al. (2011) studied the network of transnational corporations (TNCs) whose nodes are companies and (directed) links indicate that firm i owns some share of firm j. The authors showed that the TNCs exhibit a bow-tie structure with a very small core of nodes and an OUT component that is much bigger than the IN component. Combining topology with control ranking, they find that the small core is densely connected, and a node randomly drawn from it is the top holder with 50 percent probability (which reduces to 6 percent for the IN component). The authors also state they did not perform any attempt to reconstruct the network using local information, as this operation would be meaningless in an economic system whose links represent the share of ownership among economic actors.

More recently, van Lidth de Jeude et al. (2019b) have investigated the problem of reconstructing the bow-tie structure of the WTW and the DIN using several ERG-based models, namely the ones defined by purely local information (out- and in-degrees of nodes) as well as those including information about the membership of each node to a specific subgraph (i.e. SCC, IN, OUT). The WTW is characterized by a bow-tie structure where the OUT component is completely missing and the size of the SCC component increases over time together with the share of its reciprocated links (see Fig. 22). From an economic point of view, this can be interpreted in terms of an ongoing globalization process, fostered by an increasing number of trade agreements. The comparison of the aforementioned models through the AIC and BIC criteria (see Appendix for more details) allows concluding that the DCM is the best model to reconstruct the bow-tie structure of the WTW. This suggests that the information encoded into the degrees is enough to explain the peculiar mesoscale structure of the WTW.

Similarly, the DIN presents a bow-tie structure with three components (but without connections between the OUT and the IN component). The temporal evolution of the bow-tie structure is very informative about the ongoing structural organization of the system due to economic changes: the SCC size decreases before the 2008 crisis, while the size of the OUT and IN components appears as stable until they shrink around the crisis period. As for the WTW,

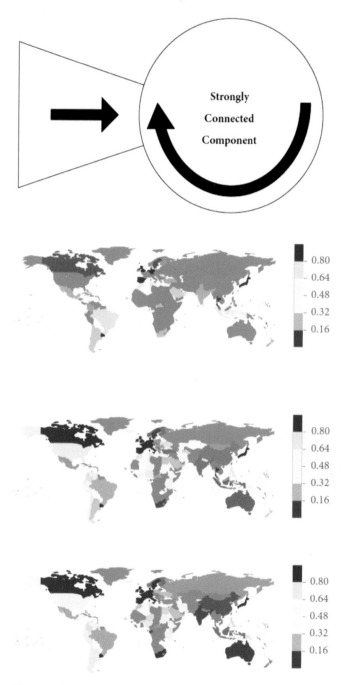

Figure 22 Bow-tie structure of the WTW. In the bottom panels, countries are colored according to their reciprocated degree (gray for countries belonging to the IN component). Source: van Lidth de Jeude et al., 2019b.

the model accounting only for the degree heterogeneity is to be preferred to block models (except in specific periods, where it competes with the RCM): therefore, both the WTW and the DIN are characterized by a peculiar bow-tie structure that, basically, seems to be explained by local constraints.

Although the bow-tie structure characterizes several metabolic, technological, and economic systems, little is known about its *dynamical* origin. In order to investigate this aspect, Zhang et al. (2007) have studied the so-called *community expertise network* (CEN), that is, the network of online questions and answers among users. Following the idea that the user answering the question should have more knowledge than the one who asked it, they built a network of expertise and found an uneven bow-tie structure with half its users belonging to the IN component (just asking questions) and almost the same amount of people ($\simeq 12\%$) belonging to the core (asking/answering) and to the OUT component (just answering questions). The authors tried to reconstruct the network's properties and its peculiar mesoscale structure by fixing a series of dynamical rules to generate a simulated network via agent-based models. They proposed two models differing in the choice of the user answering to a specific question: in the *best preferred expert* model, the probability to reply increases exponentially with the difference between the expertise of the two users (asking/answering), while in the *just better* one, the probability of answering just depends on having a slightly higher expertise than the user asking the question. They found a high similarity between the empirical bow-tie structure and the one of the network simulated via the first model, while the core of the network generated via the second model almost disappeared in favor of the tendrils.

4 Network Reconstruction at the Microscale
The Link Prediction Framework

Network reconstruction at the microscale is the task of predicting specific individual links that are missing from the network. As mentioned in the introduction, there are many cases in which this is important. Take, for instance, biological networks, such as food webs, protein–protein interaction networks, or metabolic networks (Barzel and Barabási, 2013; Cannistraci et al., 2013; Kovács et al., 2019). For these systems, whether a link between two nodes exists must be determined by field and/or laboratory experiments, which are usually very costly. Instead of blindly checking all possible interactions, focusing on those potential links that are most likely to exist can significantly reduce the experimental costs and speed the pace of uncovering the true network (Redner, 2008). Social network analysis also comes up against the missing data problem, therefore link prediction methods can be used to infer unknown relationships or collaborations (Liben-Nowell and Kleinberg, 2007). Also, the problem of *recommendation* in social networks or e-commerce applications is technically a link prediction task (Huang et al., 2005; Lü et al., 2012).

In general, besides helping in analyzing networks with missing data, link prediction algorithms can be used to guess the links that may appear in the future. From this viewpoint, any model of evolving network corresponds in principle to a link prediction algorithm (Lü et al., 2015).[12] On the other hand, link prediction techniques can be also used to identify "spurious" links resulting from inaccuracies in network data and, in general, to estimate the true structure of a network when data are noisy (Guimerà and Sales-Pardo, 2009; Newman, 2018b). For instance, protein–protein interaction networks are constructed on the basis of *measured* interactions and not of *actual* interactions, hence they can suffer not only from missing data but also from measurement errors. Another example is given by social friendship networks, which may be obtained from survey data (i.e. asking people who their friends are) that are affected by the different standards of participants on the concept of friendship itself. In these and many other cases, however, network data come with no error-assessment information of any kind (Peixoto, 2018).

The typical framework of link prediction takes as input the adjacency matrix $\mathbf{A}$ of a network, characterized by the set E of *observed links* and the set $U \setminus E$ of *nonexistent links* (U is the set of all node pairs). We assume that there are some *missing links* (or links that will appear in the future) in the set $U \setminus E$: the link prediction task is that of finding them out. Since by definition the missing links are unknown, in order to test a link prediction method the observed link set E is partitioned into a *training set* E_T and a *probe set* $E_P = E \setminus E_T$. The former is used to inform the prediction algorithm, whereas the latter is used as the prediction target. For each *nonobserved link*, namely a link in the set $E_N = E_P \cup (U \setminus E) \equiv U \setminus E_T$, the link prediction algorithm provides a score (also known as *reliability*) quantifying the likelihood of its existence. As we will see, the reliability scores can be used both to predict missing links (i.e. the nonexistent links in the observed network having the highest reliability) and to identify possible spurious links (i.e. the observed links with the lowest reliability).

Link-prediction methods can be roughly classified into two main classes: *similarity-based* and *model-based* algorithms (Lü and Zhou, 2011).[13] While similarity-based methods rely on (a varying amount of) information about the

---

[12] As such, the correct prediction of future links has a self-reinforcing feedback effect in the ability to predict further new links (Li et al., 2018).

[13] These methods are also known in the literature as *likelihood-based* or *probabilistic* methods.

network topology, model-based methods additionally assume the presence of some kind of organizing principle acting at the meso- and/or macro-scale. In both cases, the key underlying assumption is that any two nodes are more likely to interact if they have a larger *similarity*. Although "similarity" is quite an abstract concept and often depends on the specific context, it is typically proxied by the amount of direct or indirect paths between nodes (Martínez et al., 2016). In the following, for simplicity we will discuss only the case of undirected binary networks.

Similarity-Based Methods

These methods represent the simplest approach for link prediction tasks. They assign a score s_{ij} to each nonobserved link of the network that is based on the *structural similarity* of the two involved nodes i and j. This similarity is computed using solely the knowledge of the network structure around i and j, where "around" is measured in terms of *graph distance* (i.e. the number of links that form a connection path between two nodes). Similarity-based methods are then classified as local, semilocal, or global indices depending on how much the topological information used is distant from the target candidate link $i - j$.

A first remark is in order here. A crucial advantage of local and semilocal methods with respect to global ones is that they can be implemented in a decentralized manner and can, thus, handle large-scale systems. Global indices, instead, require more computational power. Typically, the computational complexity of a similarity-based method increases exponentially with the maximum network distance used, while the prediction accuracy grows sublinearly and eventually decreases. Indeed, when tested on real-world networks, global methods obtain the worst results because they make use of irrelevant information, while local techniques work surprisingly well, in turn suggesting that most of the useful information to predict links is local. However, the performance of each technique strongly depends on the structural properties of the network considered (Martínez et al., 2016).

As a second remark, let us notice that similarity-based methods should be handled with care when detecting spurious connections. As we will see, by definition these methods tend to give high scores to node pairs with many common connections, hence the *weak ties* of the network (i.e. the links connecting different regions of the network) typically get low scores. However, treating weak ties as spurious links and removing them may have the undesired effect of breaking the network into disconnected components, thus destroying its functionality. A simple solution is to adjust the similarity score with some quantity such as the *link betweenness* (the ratio of shortest paths passing through that link) (Zeng and Cimini, 2012). In this way, the weak ties that keep the network connected do not get the lowest similarity score and are not at risk of being removed.

Local Indices. Local indices consider only the information about the first neighbors of each node (i.e. the nodes at graph distance equal to 1). Denoting by Γ_i the set of neighbors of node i, with $k_i = |\Gamma_i|$ being the degree (number of neighbors) of i, the simplest index can be defined using the *preferential attachment* (PA) rule (Barabási and Albert, 1999): connection probabilities are proportional to the degree of nodes. The score assigned by this method to a nonobserved link between nodes i and j is thus:

$$s_{ij}^{[PA]} = k_i \cdot k_j. \tag{4.1}$$

A step forward is made by considering the neighborhood structure of nodes i and j. The idea is that any two nodes are more likely to be linked if they have many common neighbors (CN) (Newman, 2001; Kossinets, 2006). The popular CN index is thus

$$s_{ij}^{[CN]} = |\Gamma_i \cap \Gamma_j|; \tag{4.2}$$

note that $s_{ij}^{[CN]}$ is also equal to $(\mathbf{A}^2)_{ij}$, the number of different paths of length 2 connecting i and j. Despite its simplicity, this measure performs surprisingly well on most real-world networks and can beat much more complicated approaches. Due to its success, many variations of this recipe exist. For instance,[14] the Jaccard coefficient penalizes the nodes with many neighbors, whereas the Leicht–Holme–Newman index (LHN) (Leicht et al., 2006) compares the CN score with the expected number of common neighbors under the PA model:

$$s_{ij}^{[Jaccard]} = \frac{|\Gamma_i \cap \Gamma_j|}{k_i + k_j}, \qquad s_{ij}^{[LHN]} = \frac{|\Gamma_i \cap \Gamma_j|}{k_i \cdot k_j}. \tag{4.4}$$

More refined methods take into account the degrees of the common neighbors, thus assigning more weight to the less-connected nodes. These are the Adamic-Adar (AA) (Adamic and Adar, 2003) and the resource allocation (RA) (Zhou et al., 2009) indices:

---

[14] Other examples are the Salton index (i.e. the cosine similarity), the Sørensen index, and the Hub Promoted/Depressed index (Ravasz et al., 2002), i.e.

$$s_{ij}^{[Salton]} = \frac{|\Gamma_i \cap \Gamma_j|}{\sqrt{k_i \cdot k_j}}, \qquad s_{ij}^{[Sørensen]} = \frac{2|\Gamma_i \cap \Gamma_j|}{k_i + k_j}, \qquad s_{ij}^{[HPI]} = \frac{|\Gamma_i \cap \Gamma_j|}{\min\{k_i, k_j\}}. \tag{4.3}$$

$$s_{ij}^{[AA]} = \sum_{l \in \Gamma_i \cap \Gamma_j} \frac{1}{\ln k_l}, \qquad s_{ij}^{[RA]} = \sum_{l \in \Gamma_i \cap \Gamma_j} \frac{1}{k_l}. \qquad (4.5)$$

The local naïve Bayes model (LNB) (Liu et al., 2011), instead, gives more weight to common neighbors with higher values of the clustering coefficient (recall that the clustering coefficient c_l of a node l is the number of closed paths of length 3 involving that node, normalized by the maximum possible number of such paths, i.e. $c_l = (\mathbf{A}^3)_{ll}/[k_l(k_l - 1)]$). By assuming that the clustering coefficients of the common neighbors are mutually independent, this index can be expressed as

$$s_{ij}^{[LNB]} = \sum_{l \in \Gamma_i \cap \Gamma_j} \ln\left[\frac{c_l}{1 - c_l}\right]; \qquad (4.6)$$

notice that the different weighting procedures implemented by the AA-RA and LNB indices can be combined in a straightforward way; for instance, by multiplying term by term the corresponding sums (Liu et al., 2011).

Semilocal Indices. These indices rely on structural information up to the second neighbors of nodes i and j (i.e. the nodes at graph distance less or equal to 2). The local path (LP) index (Lü et al., 2009) is defined as

$$s_{ij}^{[LP]} = (\mathbf{A}^2 + \epsilon\mathbf{A}^3)_{ij}, \qquad (4.7)$$

where $\epsilon < 1$ is a free parameter. The LP index, thus, counts the number of paths connecting i and j of length 2 and 3, while penalizing the latter by the factor ϵ. Clearly, LP reduces to CN when $\epsilon = 0$. However, LP can also be extended to account for paths longer than three, that is,

$$s_{ij}^{[LP]} = (\mathbf{A}^2 + \epsilon\mathbf{A}^3 + \epsilon^2\mathbf{A}^4 + \cdots + \epsilon^{n-2}\mathbf{A}^n)_{ij}, \qquad (4.8)$$

where $n > 2$ is the maximal path length considered. This index asks for more information and computational resources as n increases. When $n \to \infty$, LP becomes equivalent to the Katz index (Katz, 1953) (see below).

A related set of indices that exploit the local structure of the neighborhood is given by the LCP-based methods (Cannistraci et al., 2013). These indices rely on the idea that i and j are more likely to be connected if their common neighbors are members of a strongly connected local community – an assumption known as the *local community paradigm* (LCP). The simplest LCP index is the CAR similarity, a function of both the number of common neighbors and of the number of links between them (i.e. the number of links constituting the local community):

$$s_{ij}^{[CAR]} = s_{ij}^{[CN]} \cdot \sum_{l \in \Gamma_i \cap \Gamma_j} \frac{\gamma_l}{2} = s_{ij}^{[CN]} \cdot \sum_{l \in \Gamma_i \cap \Gamma_j} \sum_{m \in \Gamma_i \cap \Gamma_j} \frac{a_{lm}}{2}, \qquad (4.9)$$

where the factor $\gamma_l = \sum_{m\in\Gamma_i\cap\Gamma_j} a_{lm}$ counts how many neighbors of node $l \in \Gamma_i \cap \Gamma_j$ are also linked to both i and j. Analogous modifications can be applied to the other similarity indices discussed above – see Cannistraci et al., 2013.

As we have seen up to now, most of the local similarity-based link prediction algorithms rely on the common neighbor idea, rooted in social network analysis, that the more common friends two individuals have, the more likely they know each other. This principle is known as *triadic closure* and leads to the key role of paths of length 2. On the other hand, semilocal indices also rely on paths of length 3, which can play a key role for link prediction in some contexts. Notably, this is the case of the protein–protein interaction (PPI) networks: proteins turn out to interact not if they are similar to each other, but if one of them is similar to the partners of the other (Kovács et al., 2019). See Fig. 23. Moreover, PPI networks were shown to display a kind of LCP architecture where protein complexes are confined in topologically isolated network structures, which are often coincident with functional network modules that play an important role in molecular circuits (Muscoloni et al., 2018). Similar observations do apply to foodweb trophic relations and world trade network transitions (Muscoloni et al., 2018). In all these cases, similarity-based link prediction metrics have to rely on the *quadrangular closure* principle (Daminelli et al., 2015). For instance, a *degree-normalized L3 score* (DnL3) has been defined in the context of PPI by Kovács et al. (2019):

$$s_{ij}^{[DnL3]} = \sum_{l\neq i,m}\sum_{m\neq j} \frac{a_{il}a_{lm}a_{mj}}{\sqrt{k_lk_m}} = \sum_{l\in\Gamma_i}\sum_{m\in\Gamma_j} \frac{a_{lm}}{\sqrt{k_lk_m}}, \tag{4.10}$$

a definition that can be extended to account for paths of length n (Muscoloni et al., 2018). Note that the role of paths of length 3 is obvious in the case of bipartite networks: since in this case connections exist only across (and not within) two sets of nodes, by definition nodes belonging to different sets can be connected only by paths of odd length (Kunegis et al., 2010) (see also the discussion at the end of the section).

Global Indices. Finally, global indices are computed by using the topological information encoded into the whole network. As mentioned above, the Katz index (Katz, 1953) is the extension of the LP index that accounts for paths of any length, that is,

$$s_{ij}^{[Katz]} = \sum_{m=1}^{\infty} \beta^m (\mathbf{A}^m)_{ij} = [(\mathbf{I} - \beta\mathbf{A})^{-1} - \mathbf{I}]_{ij}, \tag{4.11}$$

where the damping factor β suppresses the exponential proliferation of longer paths in the network. To ensure convergence, β must be smaller than the

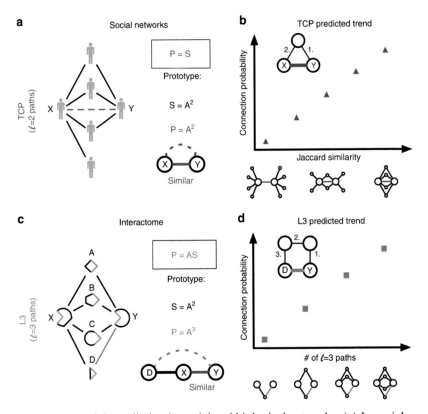

Figure 23 Link prediction in social and biological networks. **(a)** In social networks, a large number of common friends (i.e. paths of length 2) implies a higher chance to become friends (red link between nodes X and Y). This is known as the Triadic Closure Principle (TCP). TCP predicts (P) links based on node similarity (S), quantifying the number of shared neighbors between each node pair ($\mathbf{A}^2$). **(b)** TCP implies that node pairs of high Jaccard similarity (a sample index based on paths of length 2) are more likely to interact. **(c)** Protein interactions instead often require complementary interfaces. That is, two proteins X and Y sharing many neighbors will have similar interfaces, but typically this does not guarantee that X and Y directly interact with each other. Instead, Y might interact (blue link) with a neighbor of X (protein D). Such a link can be predicted through a Quadrangular Closure Principle, that is, by using paths of length 3, since these paths identify similar nodes to the known partners ($P = \mathbf{A}S = \mathbf{A}^3$). **(d)** The two proteins Y and D are more likely to interact if they are linked by multiple paths of length 3 in the network. Source: Kovács et al., 2019.

reciprocal of the largest eigenvalue of $\mathbf{A}$. Notice that the computational complexity of LP in an uncorrelated network is $O(N\langle k\rangle^n)$, converging to the complexity of the Katz index that is $O(N^3)$. Experimental evidence suggests that the optimal n is positively correlated with the average shortest distance of the network (Lü et al., 2009).

Another global index is the *local random walk* (LRW) index (Liu and Lü, 2010), defined as

$$s_{ij}^{[LRW]}(t) = \sum_{t'=1}^{t} \left[q_i \pi_{ij}(t') + q_j \pi_{ji}(t') \right], \tag{4.12}$$

where $q_i \propto k_i$ is the probability that the walker is initially on node i and $\pi_{ij}(t')$ is the probability that this walker lands at node j after t' steps. Hence, this index measures the overall probability that a walker starting on either i and j will land on the other node within t steps. Clearly, t denotes the maximum length of the considered paths. For $t \to \infty$ the index reflects the steady-state behavior of the walker and is known under the name of *random walk with restart* (RWR) index (Tong et al., 2006). Otherwise, if a small amount of resistance is added on each link of the network (and a larger amount if the link has not been traveled yet), the index becomes the *random walk with resistance* (RWS) index (Lei and Ruan, 2012).

Two related variants are the SimRank (Jeh and Widom, 2002) and the Global Leicht–Holme–Newman (GLHN) index (Leicht et al., 2006), both defined in a recursive way using the concept that two nodes are similar if their immediate neighbors are themselves similar. More quantitatively,

$$s_{ij}^{[SimRank]} = \frac{1}{k_i k_j} \sum_{l \in \Gamma_i} \sum_{m \in \Gamma_j} s_{lm}^{[SimRank]}, \qquad s_{ij}^{[GLHN]} \simeq \frac{\alpha}{\lambda_1} \sum_{l} A_{il} \frac{k_l}{k_i} s_{lj}^{[GLHN]}. \tag{4.13}$$

A different index is the Average Commute Time (ACT) index (Fouss et al., 2007), which builds on the idea that nodes i and j are similar if only a few steps are required to go from one to the other. It is defined as

$$s_{ij}^{[ACT]} = \frac{1}{l_{ii}^+ + l_{jj}^+ - 2l_{ij}^+}, \tag{4.14}$$

where the denominator is proportional to the average commute time between i and j and is expressed using the elements of the pseudoinverse of the Laplacian matrix $\mathbf{L}^+$.

The structural perturbation method (SPM) (Lü et al., 2015), instead, relies on the spectrum of the adjacency matrix $\mathbf{A}$ (i.e. the set of its eigenvalues $\{\lambda_n\}_{n=1}^N$ and normalized eigenvectors $\{\mathbf{v}_n\}_{n=1}^N$) reflecting the network's structural features. Indeed, any perturbation $\Delta\mathbf{A}$ of the network causes a shift of each eigenvalue that, at the first order, is given by the projection of the perturbation on the corresponding (unperturbed) eigenvector: $\Delta\lambda_n \simeq \mathbf{v}_n^T(\Delta\mathbf{A})\mathbf{v}_n$. The perturbed matrix can then be approximated as $\tilde{\mathbf{A}} \simeq \sum_n (\lambda_n + \Delta\lambda_n)\mathbf{v}_n\mathbf{v}_n^T$. Given that the first-order spectral shifts induced by independent perturbations are strongly correlated, the entries of the approximated perturbed matrix can be taken as link prediction scores:

$$s_{ij}^{[\text{SPM}]} = \left[\sum_n \Delta \lambda_n \mathbf{v}_n \mathbf{v}_n^T \right]_{ij}. \tag{4.15}$$

Notice that if the perturbation does not significantly change the structural features of the network, the eigenvectors of the unperturbed and perturbed matrices do not change much: if this is the case, then $\tilde{\mathbf{A}}$ and $\mathbf{A} + \Delta \mathbf{A}$ become very close. The "regularity" of a network can then be measured using the similarity (or *structural consistency*) between these two matrices, without any prior knowledge of the network's organization (Lü et al., 2015). The rationale behind SPM is then that a missing link is likely to exist if its appearance has only a small effect on the structural features of the network.

Actually, this is an important observation that applies to all link prediction methods, which are designed to echo the fundamental organization and growth rules of complex networks: the precision of a link prediction algorithm tells us the extent to which the link formation process in the network can be explained by this algorithm. Missing links are then easy to predict if their addition causes few structural changes to the network, and hard to predict otherwise (Lü et al., 2015).

Information Theoretical Methods. We dedicate this section to approaches based on information theory that are more sophisticated than standard similarity-based indices but rest on the same rationale. The mutual information (MI) index (Tan et al., 2014) is defined as

$$s_{ij}^{[\text{MI}]} = -I\left[a_{ij}|(\mathbf{A}^2)_{ij}\right] \equiv -I\left[a_{ij}\right] + I\left[a_{ij}, (\mathbf{A}^2)_{ij}\right], \tag{4.16}$$

namely the self-information of the existence of a link between nodes i and j conditional on the presence of common neighbors. An explicit expression of the MI index can be derived in the case of uncorrelated networks (Tan et al., 2014).[15]

This recipe can be easily extended to the case of multiple structural features because the values of information brought by these features are additive. For instance, the neighbor set information (NSI) index (Zhu and Xia, 2015) uses the structural information given by the common neighbors and by the links across the two neighbor sets

$$s_{ij}^{[\text{NSI}]} = -I\left[a_{ij}|(\mathbf{A}^2)_{ij}\right] - \lambda I\left[a_{ij}|(\mathbf{A}^3)_{ij}\right], \tag{4.17}$$

where λ is a hybridization parameter.

---

[15] The self-information I_i of an event i whose probability is p_i equals (minus) the logarithm of the occurrence probability of the event itself, i.e. $I_i = -\ln p_i$.

The path entropy (PE) index, instead, considers the self-information of all shortest paths between a node pair (with penalization to long paths) and can be expressed as (Xu et al., 2016)

$$s_{ij}^{[PE]} = -I \left[a_{ij} \middle| \bigcup_{m>1} (\mathbf{A}^m)_{ij} \right], \qquad (4.18)$$

where the self-information of a path is approximated by the sum of the self-information of its constituent links. The idea behind this formulation is that paths with large self-information are critical substructures for the network and greatly reduce the self-information of the link between its end nodes. Note that MI, NSI, and PE make use of an increasing amount of information, hence they can be respectively classified within the local, semilocal, and global categories.

Model-Based Methods

Methods belonging to this class build on a set of assumptions about the organizing principles of the network (see Fig. 24). The idea is to maximize the likelihood of the observed network under a given (probabilistic) parametric model, which in turn allows evaluating the likelihood of any nonexistent link. From a practical viewpoint, the computational complexity of model-based methods is typically very high (much higher than that of similarity-based methods). In addition, these methods can be very accurate in predicting links, but only when the underlying model properly describes the network and can infer links connection probabilities. Indeed, a given model-based method should represent the optimal link prediction strategy when the network is a direct realization of the model assumed (Garcia-Perez et al., 2019). Otherwise, local methods can perform much better by mimicking the network growth process and thus acting as topological learning rules (Muscoloni et al., 2017a).

Hierarchical Model. This method is based on the evidence that many real-world networks are hierarchically organized, meaning that nodes can be divided into groups, further subdivided into groups of groups, and so forth over multiple scales. Extracting such a hierarchy can then be useful to infer missing links. The hierarchical structure of a network can be represented as a dendrogram D with N leaves (corresponding to the nodes of the network) and $N-1$ internal nodes (representing the relationships among the descendant nodes in the dendrogram). The *Hierarchical Random Graph Model* (Clauset et al., 2008) assigns a probability p_r to each internal node r, so that the connection probability of a pair of nodes (leaves) is equal to $p_{r'}$, where r' is the lowest common ancestor of these two nodes. In order to find the hierarchical random

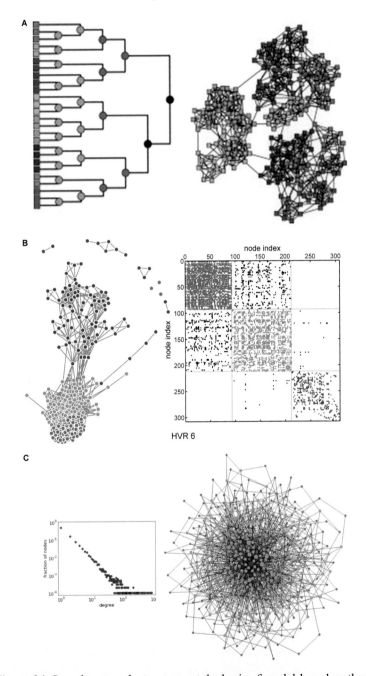

Figure 24 Sample network structures at the basis of model-based methods. (A) Dendogram and hierarchical network structure. (B) Block-structured adjacency matrix and stochastic block model network structure. (C) Scale-free degree distribution and maximum-entropy network structure obtained by constraining the degree sequence. Figure adapted from (Clauset et al., 2008; Larremore et al., 2013; Oikonomou and Cluzel, 2006).

graph $(D, \{p_r\})$ that best fits an observed real network $\mathbf{A}$, one assumes that all hierarchical random graphs are a priori equally likely. Then, the likelihood that a given model $(D, \{p_r\})$ is the correct explanation of the data is

$$\mathcal{L}(D, \{p_r\}) = \prod_r p_r^{E_r} (1 - p_r)^{L_r R_r - E_r}, \tag{4.19}$$

where E_r is the number of links whose endpoints have r as their lowest common ancestor in D, and L_r and R_r are, respectively, the number of leaves in the left and right subtrees rooted at r. Given a dendrogram D, the likelihood is maximized by the coefficients

$$p_r = \frac{E_r^*}{L_r R_r}, \ \forall \, r, \tag{4.20}$$

namely the fraction of potential links between the two subtrees of r that actually appear in the network. The Markov Chain Monte Carlo (MCMC) method can be used to sample dendrograms with probability proportional to their likelihood. Once an ensemble of dendrograms is generated, the ensemble average of the connection probability for each pair of unconnected nodes represents the prediction score of the corresponding link.

Stochastic Block Model. The SBM, which was introduced in the previous section, is an extremely popular network model: given a partition $\mathcal{M}$ of the network such that each node belongs to exactly one group, the model assumes that the connection probability for any two nodes belonging respectively to groups r and s can be indicated as p_{rs}. Hence, the likelihood of the observed network structure can be written as (Guimerà and Sales-Pardo, 2009)

$$\mathcal{L}(\mathbf{A}|\mathcal{M}) = \prod_{r \leq s} p_{rs}^{L_{rs}} (1 - p_{rs})^{N_{rs} - L_{rs}}, \tag{4.21}$$

where L_{rs} is the observed number of links between nodes in groups r and s (or within group r, in case $r = s$) and N_{rs} is the maximum number of such links in a complete graph.[16] Similarly to the previous case, the optimal set of $\{p_{rs}\}$ that maximizes the likelihood is $p_{rs} = L_{rs}^* / N_{rs}$. Under this framework, the *reliability* of a link between nodes i and j (belonging respectively to groups r and s) is

$$R_{ij} \equiv \mathcal{L}(a_{ij} = 1|\mathbf{A}) = \frac{1}{Z} \sum_{\mathcal{M}} \left(\frac{l_{rs} + 1}{N_{rs} + 2} \right) e^{-H(\mathcal{M})}, \tag{4.22}$$

---

[16] N_{rs} is the total number of node pairs $\binom{N_r}{2}$ whenever $r = s$, and $N_r N_s$ otherwise.

where $H(\mathcal{M}) = \sum_{r \leq s} \left[\ln(N_{rs} + 1) + \ln \binom{N_{rs}}{L_{rs}} \right]$ is a function of the partition and $Z = \sum_{\mathcal{M}} e^{-H(\mathcal{M})}$. As in the previous case, no prior knowledge on the true model is assumed, meaning that $p(\mathcal{M})$ is constant.

Since the number of different partitions of N elements grows faster than any finite power of N, summing over all partitions is not possible in practice. Therefore, one can employ numerical MCMC sampling (Guimerà and Sales-Pardo, 2009) or other greedy stochastic sampling strategies (Liu et al., 2013).

Maximum-Entropy Models. ERG models that reproduce the local features of the network, discussed in Section 2, can be used for link prediction as well (Parisi et al., 2018). According to the Configuration Model (CM) – built by maximizing the Shannon entropy of the network ensemble enforcing the node degrees as constraints – the likelihood of the observed network is

$$\mathcal{L}(\mathbf{A}|\vec{x}) = \prod_{i<j} p_{ij}^{a_{ij}} (1 - p_{ij})^{1-a_{ij}}, \qquad (4.23)$$

where $p_{ij} = \frac{x_i x_j}{1 + x_i x_j}$ is the probability that nodes i and j establish a connection. These connection probabilities $\{p_{ij}\}$ can be used as link prediction scores for the nonobserved links of the network. Notice also that in the limit of sparse networks, the method simplifies to $p_{ij} \simeq k_i k_j$, namely the preferential attachment method of Eq. (4.1). The entropy-based approach to link prediction has been recently extended to account for nonlinear constraints (Adriaens et al., 2020), a generalization that has been shown to outperform other competing algorithms.

Hamiltonian Models. As we have seen, the configuration model can be framed in statistical physics terms using the Hamiltonian $H(\mathbf{A}) = \sum_i \theta_i k_i(\mathbf{A})$. The probability of the network $\mathbf{A}$ thus becomes $P(\mathbf{A}|\vec{\theta}) = \frac{e^{-H(\mathbf{A}, \vec{\theta})}}{Z(\vec{\theta})}$ where $Z(\vec{\theta})$ is the partition function. Naturally, other choices of H are possible: for instance, taking inspiration from the similarity-based indices relying on the short loops in the network, one can build a Hamiltonian of the form (Pan et al., 2016)

$$H(\mathbf{A}) = - \sum_{m \leq m_c} \beta_m \ln(\mathrm{Tr}[\mathbf{A}^m]), \qquad (4.24)$$

where the m-th term of the sum counts (approximately) the number of closed walks of length m and the sum runs up to the maximum loop length m_c. The logarithm is used to rescale each term to the same magnitude, given that $\mathrm{Tr}[\mathbf{A}^m]$ grows exponentially with the leading eigenvalue of the adjacency matrix. Differently from the CM, this formulation does not admit a closed-form solution for the partition function. This is a typical situation where maximum pseudo-likelihood methods have to be used to estimate the Lagrange multipliers $\{\beta_m\}$.

In this case, one can replace the joint likelihood of the links of the network with the product over the conditional probability of each link, given the rest of the network; then, each nonobserved link is scored by the conditional probability of adding it to the network.

Hyperbolic Latent Space Models

Latent space network models assume the existence of an embedding space where the network nodes are located, such that connections are established with probabilities that decrease with the latent distance between nodes. Link prediction in this context boils down to ranking unconnected node pairs in order of increasing latent distances between them: the closer the two unlinked nodes in the latent space, the higher the probability of a missing link.

The most popular model of this kind assumes that the latent space is hyperbolic; by doing so, it can reproduce the typical emerging properties of complex networks (sparsity, self-similarity and hierarchy, scale-freeness, small-worldness, and modular structure) (Serrano et al., 2008; Krioukov et al., 2010; Papadopoulos et al., 2012; Muscoloni and Cannistraci, 2018b). In particular, the *Hyperbolic Random Graphs* (HRG) model is defined on the two-dimensional hyperbolic disk of constant negative curvature $K = -1$, so that each node i is identified by two hyperbolic coordinates: its radius r_i and angle θ_i. Therefore, the hyperbolic distance x_{ij} between two nodes i and j is given by the hyperbolic law of cosines

$$\cosh x_{ij} = \cosh r_i \cosh r_j - \sinh r_i \sinh r_j \cos \Delta\theta_{ij}, \qquad (4.25)$$

where $\Delta\theta_{ij} \simeq \theta_i - \theta_j$ is the angle between i and j. In the HRG, i and j are connected with probability

$$p\left(x_{ij}\right) = \left(1 + e^{\frac{x_{ij}-R}{2T}}\right)^{-1}, \qquad (4.26)$$

where the model parameters are the hyperbolic disk radius $R > 0$ and the temperature $T \in [0, 1)$.

Link prediction with hyperbolic geometry is a two-step procedure. The first step, *network embedding*, consists in inferring node coordinates. The typical approach consists in finding the set of coordinates that maximize the likelihood that the network has been generated by an HRG (Papadopoulos et al., 2015; Wang et al., 2016). Since node pairs are connected independently, the likelihood is given by

$$\mathcal{L}\left(a_{ij}|\{r_i, \theta_i\}, T, R\right) = \prod_{i<j} \left[p\left(x_{ij}\right)\right]^{a_{ij}} \left[1 - p\left(x_{ij}\right)\right]^{1-a_{ij}}; \qquad (4.27)$$

alternative approaches are based on Laplacian eigenmaps (Alanis-Lobato et al., 2016) and coalescent embedding (Muscoloni et al., 2017b). After coordinates have been inferred, hyperbolic distances can be used to find the most likely missing link candidates.

Hyperbolic models have been found to make good link predictions only when node coordinates are inferred accurately (Kitsak et al., 2020). In this case, they work well also if the fraction of missing links is high, and as the other model-based methods can effectively predict the nonlocal links (those between nodes that do not have any common neighbors). Link prediction performance can be further improved by computing the distances over the network topology, that is, ranking unconnected node pairs according to their hyperbolic shortest path length (the sum of the hyperbolic distances over the shortest path between these two nodes) (Muscoloni and Cannistraci, 2018a).

Network Reconstruction from Noisy Data

By postulating an underlying parametric model for the network under consideration, model-based methods have been widely used not only to predict individual missing or spurious links but also to reconstruct the more reliable network structure when the observed network data is noisy. Here we discuss the case of the SBM, which is, by far, the most popular model in the literature. Given an observed network $\mathbf{A}$, the *reliability* of an entire network $\tilde{\mathbf{A}}$ is the likelihood that $\tilde{\mathbf{A}}$ is the true network given the observation $\mathbf{A}$ (and the model belonging to the SBM family) (Guimerà and Sales-Pardo, 2009). Analogously to Eq. (4.22),

$$R(\tilde{\mathbf{A}}) \equiv \mathcal{L}(\tilde{\mathbf{A}}|\mathbf{A}) = \frac{1}{Z} \sum_{\mathcal{M}} e^{\sum_{r \leq s}\left[\ln\left(\frac{N_{rs}+1}{2N_{rs}+1}\right)+\ln\left(\frac{\binom{N_{rs}}{L_{rs}}}{\binom{2N_{rs}}{L_{rs}+\tilde{L}_{rs}}}\right)\right]} e^{-H(\mathcal{M})}. \quad (4.28)$$

Finding the network $\tilde{\mathbf{A}}$ whose reliability is maximum is, however, computationally demanding. A simple alternative, greedy algorithm consists in evaluating the link reliability for all pairs of nodes and then iteratively removing the links with lowest reliability while adding non-links with high reliability. Each move is accepted only if the total network reliability increases.

A recent generalization of this approach consists in coupling the SBM generative process with a noisy measurement model and performing Bayesian statistical inference of this joint model (Peixoto, 2018). Consider a true network $\tilde{\mathbf{A}}$ and a noisy observation of it, say $\mathbf{A}$. The inference framework to obtain $\tilde{\mathbf{A}}$ from $\mathbf{A}$ is based on:

- The network generating process. Using the SBM, a network is generated with probability

$$P(\tilde{\mathbf{A}}|\mathcal{M}, \{p_{rs}\}) = \prod_{r \le s} p_{rs}^{\tilde{L}_{rs}} (1 - p_{rs})^{\tilde{N}_{rs} - \tilde{L}_{rs}}; \qquad (4.29)$$

- The data generating process $P(\mathbf{A}|\tilde{\mathbf{A}}, \mu, \nu)$, where μ is the probability of observing a missing link (i.e. a link that exists in $\tilde{\mathbf{A}}$ but not in $\mathbf{A}$) and ν is the probability of observing a spurious link (i.e. a link that exists in $\mathbf{A}$ but not in $\tilde{\mathbf{A}}$). Without any prior knowledge, these error rates lie anywhere in the unit interval.

Under these assumptions, the final likelihood for the observed network $\mathbf{A}$ is identical to that of an effective SBM, that is,

$$P(\mathbf{A}|\mu, \nu, \mathcal{M}, \{p_{rs}\}) = \sum_{\tilde{\mathbf{A}}} P(\mathbf{A}|\tilde{\mathbf{A}}, \mu, \nu) P(\tilde{\mathbf{A}}|\mathcal{M}, \{p_{rs}\})$$

$$= \prod_{r \le s} q_{rs}^{L_{rs}} (1 - q_{rs})^{N_{rs} - L_{rs}}, \qquad (4.30)$$

with $q_{rs} = (1 - \mu - \nu)p_{rs} + \nu$ being an effective SBM-induced probability (scaled and shifted by the noise).

If the network partition $\mathcal{M}$ is known and if the number of modules is very small compared to the total number of nodes, the posterior distribution for q_{rs} should be peaked around the maximum likelihood estimate L_{rs}/N_{rs}. In this case, computing the joint posterior distribution for μ and ν returns constraints implying that the inferred error rates are bounded by the maximum and minimum inferred connection probabilities (Peixoto, 2018):

$$\hat{\nu} \le \min_{rs}\{L_{rs}/N_{rs}\}, \qquad \hat{\mu} \le 1 - \max_{rs}\{L_{rs}/N_{rs}\}; \qquad (4.31)$$

these bounds mean that μ (ν) is small if we do (do not) observe many links between groups. Hence, as long as the inferred SBM probabilities are sufficiently heterogeneous (meaning that the observed network is sufficiently structured – besides being properly described by a SBM), the inferred error rates should be contained within narrow intervals. The key observation here is that the modifications induced by the error rates uniformly affect every link and non-link; thus, with structured models we can exploit the observed correlations in the measurements to infer the underlying network (and even the error rates themselves).

Lastly, the reconstruction procedure consists of determining $\tilde{\mathbf{A}}$ from the posterior distribution $P(\tilde{\mathbf{A}}|\mathbf{A}) = \frac{P(\mathbf{A}|\tilde{\mathbf{A}})P(\tilde{\mathbf{A}})}{P(\mathbf{A})}$, which defines an ensemble of possibilities for the underlying network that incorporates the amount of uncertainty resulting from the measurement. However, the marginal network probability $P(\tilde{\mathbf{A}}) = \sum_{\mathcal{M}} P(\tilde{\mathbf{A}}|\mathcal{M})P(\mathcal{M})$ involves an intractable sum over all possible network partitions; hence, instead of directly computing the posterior, one computes the joint posterior $P(\tilde{\mathbf{A}}, \mathcal{M}|\mathbf{A}) \sim P(\mathbf{A}|\tilde{\mathbf{A}})P(\tilde{\mathbf{A}}|\mathcal{M})P(\mathcal{M})$, which defines simultaneously an ensemble of possibilities for the underlying network and its large-scale hierarchical modular organization (and involves only quantities that can be computed exactly). The original posterior can then be obtained by marginalization as $P(\tilde{\mathbf{A}}|\mathbf{A}) = \sum_{\mathcal{M}} P(\tilde{\mathbf{A}}, \mathcal{M}|\mathbf{A})$ and sampled using MCMC methods.

The described setup is sufficiently general and can be used with any variant of the SBM, like the degree-corrected and the hierarchical ones, as well as different models for the noise. Additionally, this framework has been extended to situations in which several (possibly repeated) observations are available for a network (Newman, 2018b) and when explicit information on the measurement error is available (Peixoto, 2018), as well as to include data on dynamical processes taking place on the network (Peixoto, 2019).

Quality Metrics for Link Prediction

Network reconstruction at the microscale is typically assessed against its ability to correctly predict the presence of individual links (i.e. the position of 1s in the binary adjacency matrix) and their absence (i.e. the position of 0s). From this purely topological perspective, a reconstruction algorithm can be seen as a binary classifier that determines if a given pair of unconnected nodes is linked or not. Hence, the evaluation metrics are those derived from the *confusion matrix* (see Fawcett [2006] for an exhaustive treatment of the topic). Given a "true" binary matrix $\mathbf{A}$ and the reconstructed one $\hat{\mathbf{A}}$, for each pair of nodes we have four possible combinations:

- $a_{ij} = 1$ and $\hat{a}_{ij} = 1$: an existing link has been correctly predicted and we have a *true positive*;
- $a_{ij} = 1$ but $\hat{a}_{ij} = 0$: an existing link has been incorrectly predicted as missing and we have a *false negative*;
- $a_{ij} = 0$ but $\hat{a}_{ij} = 1$: a missing link has been incorrectly predicted as existing and we have a *false positive*;
- $a_{ij} = 0$ and $\hat{a}_{ij} = 0$: a missing link has been correctly predicted and we have a *true negative*.

The total number of events within these four categories, labeled respectively as TP, FN, FP, and TN, are used to define various performance metrics, such as:

- the *sensitivity* (*true positive rate*) or *recall*, that is, the fraction of existing links that are correctly recovered:

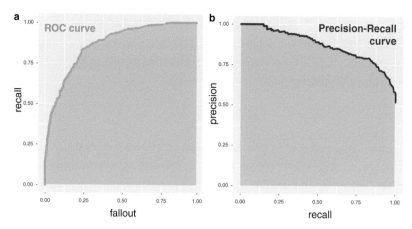

Figure 25 (a) Example of ROC curve, defined by plotting the *recall* score (or *true positive rate*) versus the *fallout* score (or *false positive rate*). The grey region denotes the *area under the ROC curve* (AUC) (b) Example of *precision-recall* curve, with the *precision* score on the vertical axis and the *recall* score on the horizontal axis. The grey region denotes the *area under the precision-recall curve* (AUPR). Source: Chicco, 2017.

$$\text{TPR} = \frac{\text{TP}}{\text{TP} + \text{FN}}; \qquad (4.32)$$

- the *specificity* (*true negative rate*), the fraction of nonexisting links that are correctly recovered:

$$\text{TNR} = \frac{\text{TN}}{\text{TN} + \text{FP}}. \qquad (4.33)$$

A good classifier should be able to achieve high values for both these quantities. Unfortunately, this may be rather difficult. Indeed, a link prediction method characterized by a high discrimination threshold (meaning that only the candidate links with the largest reliability score are predicted as actually missing) likely generates many false negatives, thus achieving a low TPR but a large TNR. Instead, if the discrimination threshold is low, the method likely generates many false positives, thus achieving a low TNR (i.e. high *false positive rate* or *fallout*, defined as FPR $= 1 - $ TNR) but a large TPR. Plotting the TPR against the FPR (i.e. *recall* VS *fallout*) as the discrimination threshold is varied generates the ROC (*Receiver Operating Characteristic*) curve, a graphical way to illustrate the performance of a binary classifier (see Fig. 25). Under this representation, the point $(0, 1)$ corresponds to the perfect classifier, yielding neither false positives nor false negatives, whereas a random classifier lies along the line of no-discrimination, namely the diagonal between $(0, 0)$ and $(1, 1)$. The *area under the ROC curve* (AUC) has been largely used to quantify the overall performance of a classifier.

An equivalent formulation of AUC is the probability that the classifier ranks a random missing link (i.e. a link in E_P) higher than a random nonexisting one (i.e. a link in $U \setminus E$). This is equivalent to performing the *Mann-Whitney U test* computing the quantity

$$\text{AUC} = \frac{n' + \frac{1}{2}n''}{n}, \tag{4.34}$$

where n is the number of times a missing link gets a higher score than a nonexistent one, n'' is the number of ties in this comparison and n is the total number of comparisons (i.e. the number of missing links times the number of nonexistent links). One achieves the result $\text{AUC} = 1/2$ if all links to be predicted get the same score, and $\text{AUC} = 1$ if all missing links occupy the top positions of the ranking.

The use of AUC has, however, been questioned since in link prediction applications it is common to deal with imbalanced network data, where one class (missing links) is overrepresented with respect to the other (existing links). This happens because most real networks are sparse, with a number of links L that is proportional to the number of nodes N rather than to the number of node pairs, namely $O(N^2)$. Hence, when testing a link prediction scheme, the set of nonobserved links will be dominated by the set of nonexisting links (set $U \setminus E$). The situation is even more intricate because a nonexisting link can result either from measuring the absence of the interaction or from not measuring the interaction at all, and typically network data come with no distinction between these two cases (Peixoto, 2018).

In any case, the result of the class imbalance is that correctly guessing nonexisting links (the true negative instances) is far easier than correctly guessing existing links (the true positives) simply because of their absolute numbers. As such, weighting sensitivity and specificity equally (as AUC does) can lead to misleading conclusions on the performance of a link prediction method. A possible solution consists in replacing the ROC curve with the *precision-recall* curve (Yang et al., 2015), where the *precision* (or *positive predicted value*) is defined as $\text{PPV} = \frac{\text{TP}}{\text{TP}+\text{FP}}$ (see Fig. 25). This is because the optimization of the ROC curve tends to maximize both the TP (through the *recall*) and the TN (through the *fallout*), whereas the optimization of the precision-recall curve tends to maximize only the TP (through both the *recall* and the *precision*) without directly considering the TN (absent in both formulas). An alternative, recently proposed solution consists in employing the *Matthews correlation coefficient*

$$\text{MCC} = \frac{\text{TP} \cdot \text{TN} - \text{FP} \cdot \text{FN}}{\sqrt{(\text{TP} + \text{FP}) \cdot (\text{TP} + \text{FN}) \cdot (\text{TN} + \text{FP}) \cdot (\text{TN} + \text{FN})}}, \tag{4.35}$$

which correctly takes into account the size of the confusion matrix elements (Chicco, 2017; Chicco and Jurman, 2020).

5 Conclusions

When studying social, economic, and biological systems, one often has access to limited information about the structure of the underlying network. The need to compensate for the scarcity of data has led to the birth of a research field known as *network reconstruction*. In the previous sections we have discussed techniques to reconstruct networks at the macro-, meso-, and micro-scale by either *constraining* or *targeting* very general network structures. We now provide a brief overview of recent research in the field that has remained outside this dissertation as well as of future perspectives.

Matrix Completion. A problem that is closely related to network reconstruction and link prediction is that of matrix completion: *given a partially observed matrix* $\mathbf{W}$*, under which conditions is it possible to recover the whole matrix exactly?* While the problem is, by definition, underdetermined, much of the research in computational mathematics focuses on the case of low-rank matrices – where the rank r is the number of linearly independent matrix columns. This case is of interest when a few variables are assumed to explain the structure of the matrix (e.g. in user-item ratings data, user preferences can often be described by a few factors). For a low-rank matrix, the number of independent entries is $O(Nr)$, where N is the dimension of $\mathbf{W}$, much smaller than the total number N^2. Thus, in this case a small number of observed elements could be sufficient for recovering it. Formally, the matrix completion problem can be stated as follows: finding the lowest rank matrix $\mathbf{X}$ that matches $\mathbf{W}$ over the set E of observed links:

$$\min_{\mathbf{X}} \text{rank}(\mathbf{X}) \quad \text{subject to } x_{ij} = w_{ij}, \ \forall \ i,j \in E. \quad (5.1)$$

While this is an NP-hard optimization problem, under proper assumptions on the sampling of the observed entries – and sufficiently many sampled entries – it can be proven to admit a unique solution, with high probability (Candes and Recht, 2009). A typical assumption is that the set of observed entries is sampled randomly – for instance, that each entry is observed with equal probability (Candes and Plan, 2010). Another important assumption concerns the coherence of nonzero elements, which should be homogeneously distributed over the whole matrix. The minimum number of observed entities for the problem to be solvable has been estimated as of the order $O(nr \log n)$ (Candes and Tao, 2010; Xu, 2018). Various matrix completion algorithms have been proposed, including convex relaxation-based algorithm (Candes and Recht, 2009), gradient-based algorithm (Keshavan et al., 2010), and alternating minimization-based algorithm (Jain et al., 2013) – the approach varies depending on whether the rank of the matrix is known or not. We refer the interested reader to Nguyen et al. (2019) for a recent, accessible review of the field.

Link Prediction for Directed and Weighted Networks. Specifically for the link prediction topic, directed networks are not easy to deal with. Since in this case paths do depend on links directionality, the simple triadic closure principle at the basis of the local similarity-based indices has to be modified to take into account the whole family of triadic motifs (Alon, 2007). Hence, for instance, the likelihood of a link $i \rightarrow j$ will be, in general, different from the likelihood of $j \rightarrow i$. The case of weighted networks is even more complicated, since no clear indication exists on how link weights should contribute to the similarity or likelihood metrics. Additionally, contradictory results have been obtained on whether the strong ties (Murata and Moriyasu, 2007) or the weak ties (Lü and Zhou, 2010) are more important to successfully predict links. An even harder problem is to predict the weights of the nonobserved links – a simple solution being to set weights of missing links proportional to their similarity scores (Zhao et al., 2015). In general, there are just a few techniques that generalize to these two more complex scenarios – notable cases being the SBM (Aicher et al., 2014) and the ERG-based models (Parisi et al., 2018).

Reconstructing Valued Networks. A big challenge is the network reconstruction task for *valued* networks, where links can have a *categorical* meaning. An example is provided by signed social networks where links can be positive or negative: in this case, network reconstruction methods have to take into account the structural balance theory (which, for instance, implies that *the enemy of my enemy is my friend*) (Leskovec et al., 2010), thus requiring additional constraints to the local ones we discussed.

Reconstructing Temporal Networks. In a different direction, the temporal evolution of links occurrences can be used to deal with the network reconstruction of temporal networks. Attempts have been done to encode the link persistence as a constraint, in order to explain the bursts of activity characterizing social networks. On the side of single link reconstruction, the evidence that older events (links) are, in general, less relevant to future links than recent ones can be directly incorporated into the similarity indices (Tylenda et al., 2009). Additionally, algorithms may benefit from the fact that in temporal networks, nodes attract links depending not only on their structural importance but also on their current level of activity (Wang et al., 2017). Moreover, when temporal data have varying periodic patterns, tensor-based techniques turn out to be particularly effective (Dunlavy et al., 2011).

Reconstructing Multiplex Networks. In multiplex networks the same set of nodes have different interaction patterns across various layers (like users interacting over multiple social networking platforms). Some attempts have been made to extend the Exponential Random Graph framework to multiplex networks: the resulting null models, however, are based on independent layers (i.e. are factorized, the factors representing single-layer null models), a characteristic that makes them suitable to be used as benchmarks and not as proper reconstruction models (a notable exception is Menichetti et al. [2014]). For what concerns the link prediction topic, algorithms based on *meta-paths* (i.e. paths across the various layers) can be used to predict links (Sun et al., 2012; Jalili et al., 2017), since the likelihood of a link increases when the ending nodes have high neighborhood similarity over multiple layers (Hristova et al., 2016; Hajibagheri et al., 2016). Multilayer mixed-membership SBM can be used in this case to develop model-based prediction methods (De Bacco et al., 2017).

Reconstruction beyond Networks. Very recently, the ERG framework has been extended to *simplicial complexes*, namely generalized network structures where interactions take place among more than two nodes at once. Although proper simplicial reconstruction techniques have not been developed yet, algorithms for link prediction can be implemented by extending the triadic closure principle as simplicial closure mechanisms (Benson et al., 2018).

Reconstruction beyond Shannon Entropy. As discussed in Section 2, the ERG formalism is based on the maximization of the Shannon entropy. However, it is in principle possible to employ non-Shannon functionals such as those belonging to the *Cressie–Read family of divergences* or *non-extensive functionals* such as the Renyi and the Tsallis entropy. Although of great interest from a purely mathematical perspective, solving the constrained maximization of such nonstandard functionals may be problematic, the main challenges being those of (1) properly extending the likelihood maximization principle and (2) finding a suitable procedure for sampling the ensemble induced by the chosen functional. For a complete overview on the topic, see Squartini et al. (2018).

Macroscale Reconstruction: Recent Case Studies. Recently, ERG-based reconstruction techniques have been applied in the context of cryptocurrencies, for example, for reconstructing the Bitcoin Lightning Network (BLN) representing transactions between users (Lin et al., 2020). Interestingly, for this network enforcing the out- and in-degree sequences (hence, using the DCM) is not enough to reproduce some centrality indicators: quantities like betweenness and eigenvector centralities are severely underestimated. This finding implies that the BLN is growing following an increasingly centralized fashion, thus becoming increasingly more fragile to failures and attacks (in particular, those aiming at splitting the network into separate components). The aforementioned analysis, however, concerns only the binary topological structure, and more work needs to be done on the weighted counterpart. Aside from the purely financial applications, a class of systems that has recently gained attention is that of *interfirm networks*, networks whose nodes are *firms* and whose links are *buying/selling relationships* between them – a construction following the so-called *supply chains* (Uchida, 2015; Watanabe et al., 2015; Goto et al., 2017; Ozaki et al., 2019). An even more interesting class of systems is that of bipartite bank-firm networks, where a link between a bank i and a firm j indicates that i has lent money to j: representations like these allow the effects of shocks at the interface between the financial and the economic system to be properly understood (Huang et al., 2013; Poledna et al., 2018).

Reconstructing these kinds of networks ultimately aims at identifying the minimal amount of information needed to reproduce the so-called *value chains*. This, in turn, opens up the possibility of generating realistic economic scenarios, to be used for testing the effects of disruptive economic events such as the crisis induced by the recent COVID-19 pandemic. The resilience of these reconstructed production networks could then be measured via the economic analogues of the regulatory stress tests; as their financial counterparts, they are expected to be very sensitive to the network features of the underlying system (Ramadiah et al., 2020a).

Appendix A
Reconstructing Bipartite Networks

By definition, the adjacency matrix $\mathbf{A}$ of a bipartite network features two empty diagonal blocks of dimension $N_1 \times N_1$ and $N_2 \times N_2$, where N_1 and N_2 are the number of nodes in the two sets. Hence a sufficient representation of a bipartite network is the $N_1 \times N_2$ *biadjacency matrix* $\mathbf{B}$, which is the off-diagonal block of the matrix $\mathbf{A}$. The generic element $b_{i\alpha}$ of the biadjacency matrix equals 1 if nodes i and α are connected, and 0 otherwise. Latin and Greek characters are used here to indicate the two sets forming the bipartite graph.

Such a representation allows generalizing the CM to reconstruct bipartite networks as well. In particular, it is possible to consider the ensemble of undirected, binary, bipartite networks $\mathcal{B}$ and solve the maximization problem of Eq. (2.21) under the constraints summed up by the Hamiltonian

$$H(\mathbf{B}) = \sum_i \theta_i k_i(\mathbf{B}) + \sum_\alpha \psi_\alpha h_\alpha(\mathbf{B}), \qquad (\text{A.1})$$

where $k_i(\mathbf{B}) = \sum_\alpha b_{i\alpha}$ and $h_\alpha(\mathbf{B}) = \sum_i b_{i\alpha}$ are the degrees of the nodes belonging to the two sets defining a bipartite network. The probability of a network in the ensemble is given by

$$P(\mathbf{B}) = \prod_i \prod_\alpha p_{i\alpha}^{b_{i\alpha}} (1 - p_{i\alpha})^{1-b_{i\alpha}}, \qquad (\text{A.2})$$

where $p_{i\alpha} = \frac{x_i y_\alpha}{1 + x_i y_\alpha}$ stands for the probability that a link exists between node i and node α; analogously to the CM, the parameters can be numerically determined by solving the system

$$k_i(\mathbf{B}^*) = \sum_\alpha \frac{x_i y_\alpha}{1 + x_i y_\alpha} = \langle k_i \rangle_{\text{BiCM}}, \; \forall \; i \qquad (\text{A.3})$$

$$h_\alpha(\mathbf{B}^*) = \sum_i \frac{x_i y_\alpha}{1 + x_i y_\alpha} = \langle h_\alpha \rangle_{\text{BiCM}}, \; \forall \; \alpha \qquad (\text{A.4})$$

that defines the *Bipartite Configuration Model* (BiCM) (Saracco et al., 2015). Notice that the system of equations defining the BiCM is formally analogous to that of defining the DCM. If *directed* bipartite networks are considered instead, the system of equations to be solved becomes more complicated (van Lidth de Jeude et al., 2019b).

Starting from the BiCM recipe, we can immediately define a bipartite version of the dcGM:

$$p_{i\alpha}^{\text{dcGM}} = \langle b_{i\alpha}\rangle_{\text{dcGM}} = \frac{zs_i t_\alpha}{1 + zs_i t_\alpha}, \ \forall \ i, \alpha, \tag{A.5}$$

where $s_i(\mathbf{V}) = \sum_\alpha w_{i\alpha}$ and $t_\alpha(\mathbf{V}) = \sum_i w_{i\alpha}$ are the strengths of the nodes belonging to the two sets, and $\mathbf{V}$ represents the weighted biadjacency matrix of the considered bipartite network; for what concerns the estimation of weights, the recipe reads

$$\langle w_{i\alpha}\rangle_{\text{dcGM}} = \frac{s_i t_\alpha}{W}, \ \forall \ i, \alpha. \tag{A.6}$$

The model defined by Eqs. (A.5) and (A.6) is also known as *Enhanced Capital-Asset Pricing Model* (ECAPM) (Squartini et al., 2017a).

Appendix B
Model Selection: A Quick Look at AIC and BIC

In a context such as that of network reconstruction, a tool is needed to find the model best fitting a given data set. A possible criterion is based on the concept of *information* and rests upon Fisher's idea that the best model is characterized by the smallest amount of information loss. The latter is computed via the Kullback–Leibler (KL) divergence, accounting for the information lost when the reality, f, is approximated by a model, g:

$$I(f,g) = \int f(x) \ln \left(\frac{f(x)}{g(x|\vec{\theta})} \right) dx \qquad (B.1)$$

(here for simplicity we consider just the unidimensional case). Hence, finding the best model means finding the model g minimizing $I(f,g)$. Information criteria are nothing else than estimations of the K-L information loss. Akaike (Akaike, 1974) found an estimator of $I(f,g)$ reading

$$\text{AIC} = -2\mathcal{L}(\hat{\theta}|\text{data}) + 2K, \qquad (B.2)$$

where $\mathcal{L}$ is the maximum of the log-likelihood function of model g (notice that the vector of parameters has been estimated through the available data) and K is the number of parameters defining the model itself (a quantity introduced to prevent overfitting). For two or more competing models, the optimal choice is the one minimizing AIC.[17] Based on the same idea, the Bayesian Information Criterion (BIC) provides an estimation of $I(f,g)$ reading (Schwarz *et al.*, 1978)

$$\text{BIC} = -2\mathcal{L}(\hat{\theta}|\text{data}) + K \ln n; \qquad (B.4)$$

in this case also, the optimal model is the one minimizing BIC.

---

[17] When the sample size is small compared to the number of parameters defining a model, a corrected version of AIC must be considered, reading

$$\text{AICc} = -2\mathcal{L}(\hat{\theta}|\text{data}) + 2K + \frac{2K(K+1)}{n-K-1}, \qquad (B.3)$$

where n is the sample size.

References

Acemoglu, Daron, Ozdaglar, Asuman, and Tahbaz-Salehi, Alireza. 2015. Systemic risk and stability in financial networks. *American Economic Review*, **105**(2), 564.

Adamic, Lada A., and Adar, Eytan. 2003. Friends and neighbors on the Web. *Social Networks*, **25**(3), 211–230.

Adamic, Lada A., and Glance, Natalie. 2005. The political blogosphere and the 2004 U.S. Election: Divided they blog. Pages 36–43 of: *Proceedings of the 3rd International Workshop on Link Discovery*. LinkKDD '05. New York, NY, USA: Association for Computing Machinery.

Adriaens, Florian, Mara, Alexandru, Lijffijt, Jefrey, and De Bie, Tijl. 2020. Scalable dyadic independence models with local and global constraints. *arXiv:2002.07076*.

Aicher, Christopher, Jacobs, Abigail Z., and Clauset, Aaron. 2014. Learning latent block structure in weighted networks. *Journal of Complex Networks*, **3**(2), 221–248.

Akaike, H. 1974. A new look at the statistical model identification. *IEEE Transactions on Automatic Control*, **19**(6), 716–723.

Alanis-Lobato, Gregorio, Mier, Pablo, and Andrade-Navarro, Miguel A. 2016. Efficient embedding of complex networks to hyperbolic space via their Laplacian. *Scientific Reports*, **6**(1), 30108.

Albert, Réka, and Barabási, Albert-László. 2002. Statistical mechanics of complex networks. *Reviews of Modern Physics*, **74**(1), 47–97.

Almeida-Neto, Mário, Guimarães, Paulo, Guimarães, Paulo R., Loyola, Rafael D., and Ulrich, Werner. 2008. A consistent metric for nestedness analysis in ecological systems: reconciling concept and measurement. *Oikos*, **117**(8), 1227–1239.

Almog, Assaf, Squartini, Tiziano, and Garlaschelli, Diego. 2017. The double role of GDP in shaping the structure of the International Trade Network. *International Journal of Computational Economics and Econometrics*, **7**(4), 381–398.

Almog, Assaf, Bird, Rhys, and Garlaschelli, Diego. 2019. Enhanced gravity model of trade: Reconciling macroeconomic and network models. *Frontiers in Physics*, **7**, 55.

Alon, Uri. 2007. Network motifs: Theory and experimental approaches. *Nature Reviews Genetics*, **8**, 450–461.

Anand, Kartik, Craig, Ben, and von Peter, Goetz. 2015. Filling in the blanks: Network structure and interbank contagion. *Quantitative Finance*, **15**(4), 625–636.

Anand, Kartik, van Lelyveld, Iman, Banai, Ádám, Friedrich, Soeren, Garratt, Rodney, Halaj, Grzegorz, Fique, Jose, Hansen, Ib, Jaramillo, Serafín Martínez, Lee, Hwayun, Molina-Borboa, José Luis, Nobili, Stefano, Rajan, Sriram, Salakhova, Dilyara, Silva, Thiago Christiano, Silvestri, Laura, and de Souza, Sergio Rubens Stancato. 2018. The missing links: A global study on uncovering financial network structures from partial data. *Journal of Financial Stability*, **35**, 107–119.

Annibale, A., Coolen, A. C. C., Fernandes, L. P., Fraternali, F., and Kleinjung, J. 2009. Tailored graph ensembles as proxies or null models for real networks I: tools for quantifying structure. *Journal of Physics A: Mathematical and Theoretical*, **42**(48), 485001.

Arasu, Arvind, Novak, Jasmine, Tomkins, Andrew, and Tomlin, John. 2002. PageRank computation and the structure of the web: Experiments and algorithms. Pages 107–117 of: *Proceedings of the Eleventh International World Wide Web Conference, Poster Track*.

Bacharach, Michael. 1965. Estimating nonnegative matrices from marginal data. *International Economic Review*, **6**(3), 294–310.

Baker, Nick J., Kaartinen, Riikka, Roslin, Tomas, and Stouffer, Daniel B. 2015. Species' roles in food webs show fidelity across a highly variable oak forest. *Ecography*, **38**(2), 130–139.

Barabási, Albert-László. 2009. Scale-free networks: A decade and beyond. *Science*, **325**(5939), 412–413.

Barabási, Albert-László, and Albert, Reka. 1999. Emergence of scaling in random networks. *Science*, **286**(5439), 509–512.

Bardoscia, Marco, Battiston, Stefano, Caccioli, Fabio, and Caldarelli, Guido. 2015. DebtRank: A microscopic foundation for shock propagation. *PLoS ONE*, **10**(6), e0130406.

Bardoscia, Marco, Battiston, Stefano, Caccioli, Fabio, and Caldarelli, Guido. 2017. Pathways towards instability in financial networks. *Nature Communications*, **8**, 14416.

Barucca, Paolo, and Lillo, Fabrizio. 2016. Disentangling bipartite and core-periphery structure in financial networks. *Chaos, Solitons & Fractals*, **88**, 244–253.

Barucca, Paolo, and Lillo, Fabrizio. 2018. The organization of the interbank network and how ECB unconventional measures affected the e-MID overnight market. *Computational Management Science*, **15**(1), 33–53.

Barucca, Paolo, Caldarelli, Guido, and Squartini, Tiziano. 2018. Tackling information asymmetry in networks: A new entropy-based ranking index. *Journal of Statistical Physics*, **173**(3), 1028–1044.

Barzel, Baruch, and Barabási, Albert-László. 2013. Network link prediction by global silencing of indirect correlations. *Nature Biotechnology*, **31**(8), 720–725.

Bascompte, Jordi, Jordano, Pedro, Melián, Carlos J., and Olesen, Jens M. 2003. The nested assembly of plant-animal mutualistic networks. *Proceedings of the National Academy of Sciences*, **100**(16), 9383–9387.

Bassett, Danielle S., Wymbs, Nicholas F., Rombach, M. Puck, Porter, Mason A., Mucha, Peter J., and Grafton, Scott T. 2013. Task-based core-periphery organization of human brain dynamics. *PLoS Computational Biology*, **9**(9), e1003171.

Battiston, Stefano, Puliga, Michelangelo, Kaushik, Rahul, Tasca, Paolo, and Caldarelli, Guido. 2012. DebtRank: Too central to fail? Financial networks, the FED and systemic risk. *Scientific Reports*, **2**, 541.

Battiston, Stefano, Farmer, J. Doyne, Flache, Andreas, Garlaschelli, Diego, Haldane, Andrew G., Heesterbeek, Hans, Hommes, Cars, Jaeger, Carlo, May, Robert, and Scheffer, Marten. 2016. Complexity theory and financial regulation. *Science*, **351**(6275), 818–819.

Benson, Austin R., Abebe, Rediet, Schaub, Michael T., Jadbabaie, Ali, and Kleinberg, Jon. 2018. Simplicial closure and higher-order link prediction. *Proceedings of the National Academy of Sciences*, **115**(48), E11221–E11230.

Borgatti, Stephen P., and Everett, Martin G. 2000. Models of core/periphery structures. *Social Networks*, **21**(4), 375–395.

Boyd, John P., Fitzgerald, William J., and Beck, Robert J. 2006. Computing core/periphery structures and permutation tests for social relations data. *Social Networks*, **28**(2), 165–178.

Broder, Andrei, Kumar, Ravi, Maghoul, Farzin, Raghavan, Prabhakar, Rajagopalan, Sridhar, Stata, Raymie, Tomkins, Andrew, and Wiener, Janet. 2000. Graph structure in the web. *Computer Networks*, **33**(1–6), 309–320.

Bruckner, Sharon, Hüffner, Falk, and Komusiewicz, Christian. 2015. A graph modification approach for finding core–periphery structures in protein interaction networks. *Algorithms for Molecular Biology*, **10**(1), 16.

Bruno, Matteo, Saracco, Fabio, Garlaschelli, Diego, Tessone, Claudio J., and Caldarelli, Guido. 2020. The ambiguity of nestedness under soft and hard constraints. *Scientific Reports* **10**, 19903.

Caccioli, Fabio, Shrestha, Munik, Moore, Cristopher, and Farmer, J. Doyne. 2014. Stability analysis of financial contagion due to overlapping portfolios. *Journal of Banking & Finance*, **46**, 233–245.

Caldarelli, G., Capocci, A., De Los Rios, P., and Muñoz, M.A. 2002. Scale-free networks from varying vertex intrinsic fitness. *Physical Review Letters*, **89**(25), 258702.

Caldarelli, Guido. 2007. *Scale-Free Networks: Complex Webs in Nature and Technology*. Oxford University Press.

Candes, E. J., and Plan, Y. 2010. Matrix Completion with noise. *Proceedings of the IEEE*, **98**(6), 925–936.

Candes, E. J., and Tao, T. 2010. The power of convex relaxation: Near-optimal Matrix Completion. *IEEE Transactions on Information Theory*, **56**(5), 2053–2080.

Candes, Emmanuel J., and Recht, Benjamin. 2009. Exact Matrix Completion via convex optimization. *Foundations of Computational Mathematics*, **9**(6), 717.

Cannistraci, Carlo Vittorio, Alanis-Lobato, Gregorio, and Ravasi, Timothy. 2013. From link-prediction in brain connectomes and protein interactomes to the local-community-paradigm in complex networks. *Scientific Reports*, **3**, 1613.

Chen, Jingchun, and Yuan, Bo. 2006. Detecting functional modules in the yeast protein–protein interaction network. *Bioinformatics*, **22**(18), 2283–2290.

Chen, Lina, Qu, Xiaoli, Cao, Mushui, Zhou, Yanyan, Li, Wan, Liang, Binhua, Li, Weiguo, He, Weiming, Feng, Chenchen, Jia, Xu, et al. 2013. Identification of breast cancer patients based on human signaling network motifs. *Scientific Reports*, **3**, 3368.

Chicco, Davide. 2017. Ten quick tips for machine learning in computational biology. *BioData Mining*, **10**, 35.

Chicco, Davide, and Jurman, Giuseppe. 2020. The advantages of the Matthews correlation coefficient (MCC) over F1 score and accuracy in binary classification evaluation. *BMC Genomics*, **21**, 6.

Chung, Fan, and Lu, Linyuan. 2002. Connected components in random graphs with given expected degree sequences. *Annals of Combinatorics*, **6**(2), 125–145.

Cimini, Giulio, Squartini, Tiziano, Gabrielli, Andrea, and Garlaschelli, Diego. 2015a. Estimating topological properties of weighted networks from limited information. *Physical Review E*, **92**(4), 040802.

Cimini, Giulio, Squartini, Tiziano, Garlaschelli, Diego, and Gabrielli, Andrea. 2015b. Systemic risk analysis on reconstructed economic and financial networks. *Scientific Reports*, **5**, 15758.

Cimini, Giulio, Squartini, Tiziano, Saracco, Fabio, Garlaschelli, Diego, Gabrielli, Andrea, and Caldarelli, Guido. 2019. The statistical physics of real-world networks. *Nature Reviews Physics*, **1**, 58–71.

Clauset, Aaron, Moore, Cristopher, and Newman, Mark E. J. 2008. Hierarchical structure and the prediction of missing links in networks. *Nature*, **453**, 98–101.

Cloutier, Mathieu, and Wang, Edwin. 2011. Dynamic modeling and analysis of cancer cellular network motifs. *Integrative Biology*, **3**(7), 724–732.

Cont, Rama, and Wagalath, Lakshithe. 2016. Fire sales forensics: Measuring endogenous risk. *Mathematical Finance*, **26**(4), 835–866.

Coolen, Antoon C. C., De Martino, Andrea, and Annibale, Alessia. 2009. Constrained Markovian dynamics of random graphs. *Journal of Statistical Physics*, **136**(6), 1035–1067.

Cover, Thomas M., and Thomas, Joy A. 2006. *Elements of Information Theory*. Wiley-Interscience.

Csermely, Peter. 2008. Creative elements: network-based predictions of active centres in proteins and cellular and social networks. *Trends in Biochemical Sciences*, **33**(12), 569–576.

Csermely, Peter, London, András, Wu, Ling-Yun, and Uzzi, Brian. 2013. Structure and dynamics of core/periphery networks. *Journal of Complex Networks*, **1**(2), 93–123.

Daminelli, Simone, Thomas, Josephine Maria, Durán, Claudio, and Cannistraci, Carlo Vittorio. 2015. Common neighbours and the local-community-paradigm for topological link prediction in bipartite networks. *New Journal of Physics*, **17**(11), 113037.

De Bacco, Caterina, Power, Eleanor A., Larremore, Daniel B., and Moore, Cristopher. 2017. Community detection, link prediction, and layer interdependence in multilayer networks. *Physical Review E*, **95**, 042317.

Della Rossa, Fabio, Dercole, Fabio, and Piccardi, Carlo. 2013. Profiling core-periphery network structure by random walkers. *Scientific Reports*, **3**, 1467.

Deming, W. Edwards, and Stephan, Frederick F. 1940. On least square adjustment of sampled frequency tables when the expected marginal totals are known. *Annals of Mathematical Statistics*, **11**(4), 427–444.

Di Gangi, Domenico, Lillo, Fabrizio, and Pirino, Davide. 2018. Assessing systemic risk due to fire sales spillover through maximum entropy network reconstruction. *Journal of Economic Dynamics and Control*, **94**, 117–141.

Dill, Stephen, Kumar, Ravi, McCurley, Kevin S., Rajagopalan, Sridhar, Sivakumar, Daksh, and Tomkins, Andrew. 2002. Self-similarity in the web. *ACM Transactions on Internet Technology (TOIT)*, **2**(3), 205–223.

Doreian, Patrick. 1985. Structural equivalence in a psychology journal network. *Journal of the American Society for Information Science*, **36**(6), 411–417.

Drehmann, Mathias, and Tarashev, Nikola. 2013. Measuring the systemic importance of interconnected banks. *Journal of Financial Intermediation*, **22**(4), 586–607.

Dueñas, Marco, Mastrandrea, Rossana, Barigozzi, Matteo, and Fagiolo, Giorgio. 2017. Spatio-temporal patterns of the international merger and acquisition network. *Scientific Reports*, **7**(1), 10789.

Dunlavy, Daniel M., Kolda, Tamara G., and Acar, Evrim. 2011. Temporal link prediction using matrix and tensor factorizations. *ACM Trans. Knowl. Discov. Data*, **5**(2).

Erdös, Paul, and Rényi, Alfred. 1960. On the evolution of random graphs. *Publications of the Mathematical Institute of the Hungarian Academy of Sciences*, **5**, 17–61.

Estrada, Ernesto, and Knight, Philip A. 2015. *A First Course in Network Theory*. Oxford University Press.

Everett, Martin G., and Borgatti, Stephen P. 1999. The centrality of groups and classes. *The Journal of Mathematical Sociology*, **23**(3), 181–201.

Faskowitz, Joshua, Yan, Xiaoran, Zuo, Xi-Nian, and Sporns, Olaf. 2018. Weighted stochastic block models of the human connectome across the life span. *Scientific Reports*, **8**(1), 12997.

Fawcett, Tom. 2006. An introduction to ROC analysis. *Pattern Recognition Letters*, **27**, 861–874.

Fienberg, Stephen E. 1970. An iterative procedure for estimation in contingency tables. *Annals of Mathematical Statistics*, **41**(3), 907–917.

Fienberg, Stephen E., and Wasserman, Stanley S. 1981. Categorical data analysis of single sociometric relations. *Sociological Methodology*, **12**, 156–192.

Fortunato, Santo. 2010. Community detection in graphs. *Physics Reports*, **486**(3–5), 75–174.

Fortunato, Santo, and Hric, Darko. 2016. Community detection in networks: A user guide. *Physics Reports*, **659**, 1–44.

Fouss, F., Pirotte, A., Renders, J., and Saerens, M. 2007. Random-walk computation of similarities between nodes of a graph with application to collaborative recommendation. *IEEE Transactions on Knowledge and Data Engineering*, **19**(3), 355–369.

Fricke, Daniel, and Lux, Thomas. 2015. Core–periphery structure in the overnight money market: evidence from the e-mid trading platform. *Computational Economics*, **45**(3), 359–395.

Fronczak, Piotr, Fronczak, Agata, and Bujok, Maksymilian. 2013. Exponential random graph models for networks with community structure. *Physical Review E*, **88**(3), 32810.

Furfine, Craig H. 2003. Interbank exposures: Quantifying the risk of contagion. *Journal of Money, Credit and Banking*, **35**(1), 111–128.

Gabrielli, Andrea, Mastrandrea, Rossana, Caldarelli, Guido, and Cimini, Giulio. 2019. Grand canonical ensemble of weighted networks. *Physical Review E*, **99**, 030301.

Gai, Praganna, and Kapadia, Sujit. 2010. Contagion in financial networks. *Proceedings of the Royal Society A: Mathematical, Physical and Engineering Sciences*, **466**(2120), 2401–2423.

Garcia-Perez, Guillermo, Aliakbarisani, Roya, Ghasemi, Abdorasoul, and Serrano, M. Angeles. 2019. *Predictability of Missing Links in Complex Networks*. https://arxiv.org/abs/1902.00035.

Garlaschelli, Diego, and Loffredo, Maria. 2004a. Fitness-dependent topological properties of the World Trade Web. *Physical Review Letters*, **93**(18), 1–4.

Garlaschelli, Diego, and Loffredo, Maria I. 2004b. Patterns of link reciprocity in directed networks. *Physical Review Letters*, **93**(26), 268701.

Garlaschelli, Diego, and Loffredo, Maria I. 2006. Multispecies grand-canonical models for networks with reciprocity. *Physical Review E*, **73**, 015101.

Garlaschelli, Diego, and Loffredo, Maria I. 2009. Generalized Bose-Fermi statistics and structural correlations in weighted networks. *Physical Review Letters*, **102**, 038701.

Goto, Hayato, Takayasu, Hideki, and Takayasu, Misako. 2017. Estimating risk propagation between interacting firms on inter-firm complex network. *PLoS ONE*, **12**, e0185712.

Greenwood, Robin, Landier, Augustin, and Thesmar, David. 2015. Vulnerable banks. *Journal of Financial Economics*, **115**(3), 471–485.

Gualdi, Stanislao, Cimini, Giulio, Primicerio, Kevin, Clemente, Riccardo Di, and Challet, Damien. 2016. Statistically validated network of portfolio overlaps and systemic risk. *Scientific Reports*, **6**, 39467.

Guimerà, Roger, and Amaral, Luís A. Nunes. 2005. Functional cartography of complex metabolic networks. *Nature*, **433**(7028), 895.

Guimerà, Roger, and Sales-Pardo, Marta. 2009. Missing and spurious interactions and the reconstruction of complex networks. *Proceedings of the National Academy of Sciences*, **106**(52), 22073–22078.

Hajibagheri, A., Sukthankar, G., and Lakkaraju, K. 2016. A holistic approach for predicting links in coevolving multiplex networks. Pages 1079–1086 of:

2016 IEEE/ACM International Conference on Advances in Social Networks Analysis and Mining (ASONAM).

Halaj, Grzegorz, and Kok, Christoffer. 2013. Assessing interbank contagion using simulated networks. *Computational Management Science*, **10**(2), 157–186.

Haldane, Andrew G., and May, Robert M. 2011. Systemic risk in banking ecosystems. *Nature*, **469**(7330), 351–355.

Hirate, Yu, Kato, Shin, and Yamana, Hayato. 2008. Web structure in 2005. Pages 36–46 of: Aiello, William, Broder, Andrei, Janssen, Jeannette, and Milios, Evangelos (eds.), *Algorithms and Models for the Web-Graph*. Berlin, Heidelberg: Springer.

Holland, Paul W., Laskey, Kathryn Blackmond, and Leinhardt, Samuel. 1983. Stochastic blockmodels: First steps. *Social Networks*, **5**(2), 109–137.

Holme, Petter. 2005. Core-periphery organization of complex networks. *Physical Review E*, **72**(4), 046111.

Hristova, Desislava, Noulas, Anastasios, Brown, Chloë, Musolesi, Mirco, and Mascolo, Cecilia. 2016. A multilayer approach to multiplexity and link prediction in online geo-social networks. *EPJ Data Science*, **5**, 24.

Huang, Xuqing, Vodenska, Irena, Havlin, Shlomo, and Stanley, H. Eugene. 2013. Cascading failures in bi-partite graphs: model for systemic risk propagation. *Scientific Reports*, **3**(1219).

Huang, Zan, Li, Xin, and Chen, Hsinchun. 2005. Link prediction approach to collaborative filtering. Pages 141–142 of: *Proceedings of the 5th ACM/IEEE-CS Joint Conference on Digital Libraries*. JCDL 05. New York, NY, USA: ACM.

Iori, Giulia, De Masi, Giulia, Precup, Ovidiu Vasile, Gabbi, Giampaolo, and Caldarelli, Guido. 2008. A network analysis of the Italian overnight money market. *Journal of Economic Dynamics and Control*, **32**(1), 259–278.

Jain, Prateek, Netrapalli, Praneeth, and Sanghavi, Sujay. 2013. Low-rank Matrix Completion using alternating minimization. Pages 665–674 of: *Proceedings of the Forty-Fifth Annual ACM Symposium on Theory of Computing*. STOC '13. New York, NY, USA: Association for Computing Machinery.

Jalili, Mahdi, Orouskhani, Yasin, Asgari, Milad, Alipourfard, Nazanin, and Perc, Matjaž. 2017. Link prediction in multiplex online social networks. *Royal Society Open Science*, **4**(2), 160863.

Jaynes, E. T. 1957. Information theory and statistical mechanics. *Physical Review*, **106**(4), 620–630.

Jeh, Glen, and Widom, Jennifer. 2002. SimRank: A measure of structural-context similarity. Pages 538–543 of: *Proceedings of the Eighth ACM SIGKDD International Conference on Knowledge Discovery and Data Mining.* KDD '02. New York, NY, USA: ACM.

Jiang, Shan, Fiore, Gaston A., Yang, Yingxiang, Ferreira Jr, Joseph, Frazzoli, Emilio, and González, Marta C. 2013. A review of urban computing for mobile phone traces: current methods, challenges and opportunities. Page 2 of: *Proceedings of the 2nd ACM SIGKDD International Workshop on Urban Computing.* ACM.

Johnson, Samuel, Domínguez-García, Virginia, and Muñoz, Miguel A. 2013. Factors determining nestedness in complex networks. *PLoS ONE*, **8**(9), e74025.

Karrer, Brian, and Newman, M. E. J. 2011. Stochastic blockmodels and community structure in networks. *Physical Review E*, **83**(1), 16107.

Kashtan, Nadav, and Alon, Uri. 2005. Spontaneous evolution of modularity and network motifs. *Proceedings of the National Academy of Sciences*, **102**(39), 13773–13778.

Katz, Leo. 1953. A new status index derived from sociometric analysis. *Psychometrika*, **18**(1), 39–43.

Keshavan, R. H., Montanari, A., and Oh, S. 2010. Matrix completion from a few entries. *IEEE Transactions on Information Theory*, **56**(6), 2980–2998.

Kitsak, Maksim, Voitalov, Ivan, and Krioukov, Dmitri. 2020. Link prediction with hyperbolic geometry. *Physical Review Research* **2**, 043113.

Kojaku, Sadamori, and Masuda, Naoki. 2017. Finding multiple core-periphery pairs in networks. *Physical Review E*, **96**(5), 052313.

Kojaku, Sadamori, and Masuda, Naoki. 2018. Core-periphery structure requires something else in the network. *New Journal of Physics*, **20**(4), 043012.

Kojaku, Sadamori, Cimini, Giulio, Caldarelli, Guido, and Masuda, Naoki. 2018. Structural changes in the interbank market across the financial crisis from multiple coreperiphery analysis. *Journal of Network Theory in Finance*, **4**(3), 33–51.

Komatsu, Takanori, and Namatame, Akira. 2012. Distributed consensus and mitigating risk propagation in evolutionary optimized networks. Pages 200–209 of: Kim, Jong-Hyun, Lee, Kangsun, Tanaka, Satoshi, and Park, Soo-Hyun (eds.), *Advanced Methods, Techniques, and Applications in Modeling and Simulation.* Tokyo: Springer Japan.

Kossinets, Gueorgi. 2006. Effects of missing data in social networks. *Social Networks*, **28**(3), 247–268.

Kovács, István A., Luck, Katja, Spirohn, Kerstin, Wang, Yang, Pollis, Carl, Schlabach, Sadie, Bian, Wenting, Kim, Dae-Kyum, Kishore, Nishka, Hao, Tong, Calderwood, Michael A., Vidal, Marc, and Barabási, Albert-Lśzló. 2019. Network-based prediction of protein interactions. *Nature Communications*, **10**, 1240.

Krioukov, Dmitri, Papadopoulos, Fragkiskos, Kitsak, Maksim, Vahdat, Amin, and Boguñá, Marián. 2010. Hyperbolic geometry of complex networks. *Physical Review E*, **82**, 036106.

Kunegis, Jérôme, De Luca, Ernesto W., and Albayrak, Sahin. 2010. The link prediction problem in bipartite networks. Pages 380–389 of: Hüllermeier, Eyke, Kruse, Rudolf, and Hoffmann, Frank (eds.), *Computational Intelligence for Knowledge-Based Systems Design*. Springer Berlin Heidelberg.

Larremore, Daniel B., Clauset, Aaron, and Buckee, Caroline O. 2013. A network approach to analyzing highly recombinant malaria parasite genes. *PLoS Computational Biology*, **9**(10), e1003268.

Laumann, Edward O, and Pappi, Franz U. 1976. *Networks of collective action: A perspective on community influence systems*. Academic Press, New York.

Lebacher, Michael, Cook, Samantha, Klein, Nadja, and Kauermann, Göran. 2019. *In Search of Lost Edges: A Case Study on Reconstructing Financial Networks*. https://arxiv.org/abs/1909.01274.

Lee, Sang Hoon, Cucuringu, Mihai, and Porter, Mason A. 2014. Density-based and transport-based core-periphery structures in networks. *Physical Review E*, **89**(3), 032810.

Lee, Tong Ihn, Rinaldi, Nicola J., Robert, François, Odom, Duncan T., Bar-Joseph, Ziv, Gerber, Georg K., Hannett, Nancy M., Harbison, Christopher T., Thompson, Craig M., Simon, Itamar, et al. 2002. Transcriptional regulatory networks in Saccharomyces cerevisiae. *Science*, **298**(5594), 799–804.

Lei, Chengwei, and Ruan, Jianhua. 2012. A novel link prediction algorithm for reconstructing protein-protein interaction networks by topological similarity. *Bioinformatics*, **29**(3), 355–364.

Leicht, Elizabeth A., Holme, Petter, and Newman, Mark E. J. 2006. Vertex similarity in networks. *Physical Review E*, **73**, 026120.

Leskovec, Jure, and Faloutsos, Christos. 2006. Sampling from large graphs. Page 631–636 of: *Proceedings of the 12th ACM SIGKDD International Conference on Knowledge Discovery and Data Mining*. KDD '06. New York, NY, USA: Association for Computing Machinery.

Leskovec, Jure, Huttenlocher, Daniel, and Kleinberg, Jon. 2010. Predicting positive and negative links in online social networks. Pages 641–650 of: *Proceedings of the 19th International Conference on World Wide Web*. WWW ?10. New York, NY, USA: ACM.

Li, Shibao, Huang, Junwei, Zhang, Zhigang, Liu, Jianhang, Huang, Tingpei, and Chen, Haihua. 2018. Similarity-based future common neighbors model for link prediction in complex networks. *Scientific Reports*, **8**, 17014.

Liben-Nowell, David, and Kleinberg, Jon. 2007. The link-prediction problem for social networks. *Journal of the American Society for Information Science and Technology*, **58**(7), 1019–1031.

Lim, Wendell A, Lee, Connie M, and Tang, Chao. 2013. Design principles of regulatory networks: searching for the molecular algorithms of the cell. *Molecular Cell*, **49**(2), 202–212.

Lin, Jian-Hong, Primicerio, Kevin, Squartini, Tiziano, Decker, Christian, and Tessone, Claudio J. 2020. Lightning Network: a second path towards centralisation of the Bitcoin economy. *New Journal of Physics*, **22**, 083022.

Lip, Sean Z. W. 2011. *A Fast Algorithm for the Discrete Core/Periphery Bipartitioning Problem*. https://arxiv.org/abs/1102.5511.

Liu, Weiping, and Lü, Linyuan. 2010. Link prediction based on local random walk. *Europhysics Letters*, **89**(5), 58007.

Liu, Zhen, Zhang, Qian-Ming, Lü, Linyuan, and Zhou, Tao. 2011. Link prediction in complex networks: A local naïve Bayes model. *Europhysics Letters*, **96**(4), 48007.

Liu, Zhen, He, Jia-Lin, Kapoor, Komal, and Srivastava, Jaideep. 2013. Correlations between community structure and link formation in complex networks. *PLoS ONE*, **8**(9), e72908.

Lü, Linyuan, and Zhou, Tao. 2010. Link prediction in weighted networks: The role of weak ties. *EPL (Europhysics Letters)*, **89**(1), 18001.

Lü, Linyuan, and Zhou, Tao. 2011. Link prediction in complex networks: A survey. *Physica A: Statistical Mechanics and Its Applications*, **390**(6), 1150–1170.

Lü, Linyuan, Jin, Ci-Hang, and Zhou, Tao. 2009. Similarity index based on local paths for link prediction of complex networks. *Physical Review E*, **80**, 046122.

Lü, Linyuan, Medo, Matúš, Yeung, Chi Ho, Zhang, Yi-Cheng, Zhang, Zi-Ke, and Zhou, Tao. 2012. Recommender systems. *Physics Reports*, **519**(1), 1–49.

Lü, Linyuan, Pan, Liming, Zhou, Tao, Zhang, Yi-Cheng, and Stanley, H. Eugene. 2015. Toward link predictability of complex networks. *Proceedings of the National Academy of Sciences*, **112**(8), 2325–2330.

Ma, Athen, and Mondragón, Raúl J. 2015. Rich-cores in networks. *PLoS ONE*, **10**(3), e0119678.

Mariani, Manuel Sebastian, Ren, Zhuo-Ming, Bascompte, Jordi, and Tessone, Claudio Juan. 2019. Nestedness in complex networks: Observation, emergence, and implications. *Physics Reports*, **813**, 1–90.

Martínez, Víctor, Berzal, Fernando, and Cubero, Juan-Carlos. 2016. A survey of link prediction in complex networks. *ACM Comput. Surv.*, **49**(4).

Maslov, S., and Sneppen, K. 2002. Specificity and stability in topology of protein networks. *Science*, 910.

Mastrandrea, Rossana, Squartini, Tiziano, Fagiolo, Giorgio, and Garlaschelli, Diego. 2014a. Enhanced reconstruction of weighted networks from strengths and degrees. *New Journal of Physics*, **16**(4), 043022.

Mastrandrea, Rossana, Squartini, Tiziano, Fagiolo, Giorgio, and Garlaschelli, Diego. 2014b. Reconstructing the world trade multiplex: The role of intensive and extensive biases. *Physical Review E*, **90**(6), 062804.

Mastromatteo, Iacopo, Zarinelli, Elia, and Marsili, Matteo. 2012. Reconstruction of financial networks for robust estimation of systemic risk. *Journal of Statistical Mechanics: Theory and Experiment*, **2012**(03), P03011.

Mazzarisi, Piero, and Lillo, Fabrizio. 2017. *Methods for Reconstructing Interbank Networks from Limited Information: A Comparison*. Springer International Publishing. Pages 201–215.

Menichetti, Giulia, Remondini, Daniel, and Bianconi, Ginestra. 2014. Correlations between weights and overlap in ensembles of weighted multiplex networks. *Physical Review E*, **90**, 062817.

Messé, Arnaud, Hütt, Marc-Thorsten, and Hilgetag, Claus C. 2018. Toward a theory of coactivation patterns in excitable neural networks. *PLoS Computational Biology*, **14**(4), e1006084.

Milo, Ron, Shen-Orr, Shai, Itzkovitz, Shalev, Kashtan, Nadav, Chklovskii, Dmitri, and Alon, Uri. 2002. Network motifs: simple building blocks of complex networks. *Science*, **298**(5594), 824–827.

Mistrulli, Paolo Emilio. 2011. Assessing financial contagion in the interbank market: Maximum entropy versus observed interbank lending patterns. *Journal of Banking & Finance*, **35**(5), 1114–1127.

Murata, Tsuyoshi, and Moriyasu, Sakiko. 2007. Link Prediction of Social Networks Based on Weighted Proximity Measures. Pages 85–88 of: *Proceedings of the IEEE/WIC/ACM International Conference on Web Intelligence*. WI 07. USA: IEEE Computer Society.

Muscoloni, Alessandro, and Cannistraci, Carlo Vittorio. 2018a. Leveraging the nonuniform PSO network model as a benchmark for performance evaluation in community detection and link prediction. *New Journal of Physics*, **20**(6), 063022.

Muscoloni, Alessandro, and Cannistraci, Carlo Vittorio. 2018b. A nonuniform popularity-similarity optimization (nPSO) model to efficiently generate realistic complex networks with communities. *New Journal of Physics*, **20**(5), 052002.

Muscoloni, Alessandro, Michieli, Umberto, and Cannistraci, Carlo Vittorio. 2017a. *Local-ring network automata and the impact of hyperbolic geometry in complex network link-prediction.* https://arxiv.org/abs/1707. 09496.

Muscoloni, Alessandro, Thomas, Josephine Maria, Ciucci, Sara, Bianconi, Ginestra, and Cannistraci, Carlo Vittorio. 2017b. Machine learning meets complex networks via coalescent embedding in the hyperbolic space. *Nature Communications*, **8**(1), 1615.

Muscoloni, Alessandro, Abdelhamid, Ilyes, and Cannistraci, Carlo Vittorio. 2018. Local-community network automata modelling based on length-three-paths for prediction of complex network structures in protein interactomes, food webs and more. *bioRxiv*. DOI: 10.1101/346916

Musmeci, Nicolò, Battiston, Stefano, Caldarelli, Guido, Puliga, Michelangelo, and Gabrielli, Andrea. 2013. Bootstrapping topology and systemic risk of complex network using the fitness model. *Journal of Statistical Physics*, **151**, 720–734.

Newman, M. E. J. 2018a. *Networks*. Oxford University Press.

Newman, Mark E. J. 2001. Clustering and preferential attachment in growing networks. *Physical Review E*, **64**, 025102.

Newman, Mark E. J. 2002. Assortative mixing in networks. *Physical Review Letters*, **2**(4), 1–5.

Newman, Mark E. J. 2018b. Network structure from rich but noisy data. *Nature Physics*, **14**, 542–545.

Nguyen, L. T., Kim, J., and Shim, B. 2019. Low-rank Matrix Completion: A contemporary survey. *IEEE Access*, **7**, 94215–94237.

Nicolini, Carlo, and Bifone, Angelo. 2016. Modular structure of brain functional networks: Breaking the resolution limit by Surprise. *Scientific Reports*, **6**, 19250.

Ohnishi, Takaaki, Takayasu, Hideki, and Takayasu, Misako. 2010. Network motifs in an inter-firm network. *Journal of Economic Interaction and Coordination*, **5**(2), 171–180.

Oikonomou, Panos, and Cluzel, Philippe. 2006. Effects of topology on network evolution. *Nature Physics*, **2**(8), 532–536.

Onnela, Jukka-Pekka, Saramäki, Jari, Kertész, János, and Kaski, Kimmo. 2005. Intensity and coherence of motifs in weighted complex networks. *Physical Review E*, **71**, 065103.

Opsahl, Tore, Agneessens, Filip, and Skvoretz, John. 2010. Node centrality in weighted networks: Generalizing degree and shortest paths. *Social Networks*, **32**(3), 245–251.

Orsini, Chiara, Dankulov, Marija M., Colomer-de Simón, Pol, Jamakovic, Almerima, Mahadevan, Priya, Vahdat, Amin, Bassler, Kevin E., Toroczkai, Zoltán, Boguñá, Marián, Caldarelli, Guido, Fortunato, Santo, and Krioukov, Dmitri. 2015. Quantifying randomness in real networks. *Nature Communications*, **6**, 8627.

Ozaki, Junichi, Tamura, Koutarou, Takayasu, Hideki, and Takayasu, Misako. 2019. Modeling and simulation of Japanese inter-firm network. *Artificial Life and Robotics*, **24**, 257–261.

Page, Lawrence, Brin, Sergey, Motwami, R., Winograd, Terry, and Motwani, Rajeev. 1999. *The PageRank citation ranking: Bringing order to the web.* Unpublished work. Stanford InfoLab.

Palla, Gergely, Derényi, Imre, Farkas, Illés, and Vicsek, Tamás. 2005. Uncovering the overlapping community structure of complex networks in nature and society. *Nature*, **435**(7043), 814.

Pan, Liming, Zhou, Tao, Lü, Linyuan, and Hu, Chin-Kun. 2016. Predicting missing links and identifying spurious links via likelihood analysis. *Scientific Reports*, **6**, 22955.

Papadopoulos, Fragkiskos, Kitsak, Maksim, Serrano, M. Ángeles, Boguñá, Marián, and Krioukov, Dmitri. 2012. Popularity versus similarity in growing networks. *Nature*, **489**(7417), 537–540.

Papadopoulos, Fragkiskos, Aldecoa, Rodrigo, and Krioukov, Dmitri. 2015. Network geometry inference using common neighbors. *Physical Review E*, **92**, 022807.

Parisi, Federica, Caldarelli, Guido, and Squartini, Tiziano. 2018. Entropy-based approach to missing-links prediction. *Applied Network Science*, **3**, 17.

Parisi, Federica, Squartini, Tiziano, and Garlaschelli, Diego. 2020. A faster horse on a safer trail: generalized inference for the efficient reconstruction of weighted networks. *New Journal of Physics*, **22**(5), 053053.

Park, Juyong, and Newman, M. E. J. 2004a. Solution of the two-star model of a network. *Physical Review E*, 066146.

Park, Juyong, and Newman, M. E. J. 2004b. The statistical mechanics of networks. *Physical Review E*, **70**, 66117.

Park, Juyong, and Newman, M. E. J. 2005. Solution for the properties of a clustered network. *Physical Review E*, 026136.

Payrató-Borràs, Clàudia, Hernández, Laura, and Moreno, Yamir. 2019. Breaking the spell of nestedness: The entropic origin of nestedness in mutualistic systems. *Physical Review X*, **9**, 031024.

Peixoto, Tiago P. 2017. Nonparametric Bayesian inference of the microcanonical stochastic block model. *Physical Review E*, **95**(1), 012317.

Peixoto, Tiago P. 2018. Reconstructing networks with unknown and heterogeneous errors. *Physical Review X*, **8**, 041011.

Peixoto, Tiago P. 2019. Network reconstruction and community detection from dynamics. *Physical Review Letters*, **123**, 128301.

Poledna, Sebastian, Hinteregger, Abraham, and Thurner, Stefan. 2018. Identifying systemically important companies by using the credit network of an entire nation. *Entropy*, **20**, 792.

Ramadiah, Amanah, Gangi, Domenico Di, Sardo, D. Ruggiero Lo, Macchiati, Valentina, Minh, Tuan Pham, Pinotti, Francesco, Wilinski, Mateusz, Barucca, Paolo, and Cimini, Giulio. 2020a. Network sensitivity of systemic risk. *Journal of Network Theory in Finance*, **5**(3), 53–72.

Ramadiah, Amanah, Caccioli, Fabio, and Fricke, Daniel. 2020b. Reconstructing and stress testing credit networks. *Journal of Economic Dynamics and Control*, **111**, 103817.

Ravasz, E., Somera, A. L., Mongru, D. A., Oltvai, Z. N., and Barabási, A.-L. 2002. Hierarchical organization of modularity in metabolic networks. *Science*, **297**(5586), 1551–1555.

Reddy, P. Krishna, Kitsuregawa, Masaru, Sreekanth, P., and Rao, S. Srinivasa. 2002. A graph based approach to extract a neighborhood customer community for collaborative filtering. Pages 188–200 of: *International Workshop on Databases in Networked Information Systems*. Springer.

Redner, Sid. 2008. Teasing out the missing links. *Nature*, **453**, 47–48.

Rombach, M. Puck, Porter, Mason A., Fowler, James H., and Mucha, Peter J. 2014. Core-periphery structure in networks. *SIAM Journal on Applied Mathematics*, **74**(1), 167–190.

Rombach, Puck, Porter, Mason A., Fowler, James H., and Mucha, Peter J. 2017. Core-periphery structure in networks (revisited). *SIAM Review*, **59**(3), 619–646.

Saracco, Fabio, Di Clemente, Riccardo, Gabrielli, Andrea, and Squartini, Tiziano. 2015. Randomizing bipartite networks: The case of the World Trade Web. *Scientific Reports*, **5**, 10595.

Saracco, Fabio, Di Clemente, Riccardo, Gabrielli, Andrea, and Squartini, Tiziano. 2016. Detecting early signs of the 2007–2008 crisis in the world trade. *Scientific Reports*, **6**, 30286.

Schneider, Christian M., Belik, Vitaly, Couronné, Thomas, Smoreda, Zbigniew, and González, Marta C. 2013. Unravelling daily human mobility motifs. *Journal of The Royal Society Interface*, **10**(84), 20130246.

Schwarz, Gideon, et al. 1978. Estimating the dimension of a model. *The Annals of Statistics*, **6**(2), 461–464.

Serrano, M. Ángeles, Krioukov, Dmitri, and Boguñá, Marián. 2008. Self-similarity of complex networks and hidden metric apaces. *Physical Review Letters*, **100**, 078701.

Shen-Orr, Shai S., Milo, Ron, Mangan, Shmoolik, and Alon, Uri. 2002. Network motifs in the transcriptional regulation network of *Escherichia coli*. *Nature Genetics*, **31**(1), 64.

Shleifer, Andrei, and Vishny, Robert. 2011. Fire sales in finance and macroeconomics. *Journal of Economic Perspectives*, **25**(1), 29–48.

Silverstein, Jack W. 1994. The spectral radii and norms of large dimensional non-central random matrices. *Communications in Statistics. Stochastic Models*, **10**(3), 525–532.

Sinatra, Roberta, Condorelli, Daniele, and Latora, Vito. 2010. Networks of motifs from sequences of symbols. *Physical Review Letters*, **105**(17), 178702.

Smith, David A., and White, Douglas R. 1992. Structure and dynamics of the global economy: Network analysis of international trade 1965–1980. *Social Forces*, **70**(4), 857–893.

Snijders, Tom A. B., and Nowicki, Krzysztof. 1997. Estimation and prediction for stochastic blockmodels for graphs with latent block structure. *Journal of Classification*, **14**(1), 75–100.

Sporns, Olaf, and Kötter, Rolf. 2004. Motifs in brain networks. *PLoS Biology*, **2**(11), e369.

Squartini, Tiziano, and Garlaschelli, Diego. 2011. Analytical maximum-likelihood method to detect patterns in real networks. *New Journal of Physics*, **13**, 083001.

Squartini, Tiziano, and Garlaschelli, Diego. 2012. Triadic motifs and dyadic self-organization in the World Trade Network. Pages 24–35 of: *International Workshop on Self-Organizing Systems*. Springer.

Squartini, Tiziano, Fagiolo, Giorgio, and Garlaschelli, Diego. 2011a. Randomizing world trade. I. A binary network analysis. *Physical Review E*, **84**(4), 046117.

Squartini, Tiziano, Fagiolo, Giorgio, and Garlaschelli, Diego. 2011b. Randomizing world trade. II. A weighted network analysis. *Physical Review E*, **84**, 046118.

Squartini, Tiziano, van Lelyveld, Iman, and Garlaschelli, Diego. 2013a. Early-warning signals of topological collapse in interbank networks. *Scientific Reports*, **3**, 3357.

Squartini, Tiziano, Picciolo, Francesco, Ruzzenenti, Franco, and Garlaschelli, Diego. 2013b. Reciprocity of weighted networks. *Scientific Reports*, **3**(1), 2729.

Squartini, Tiziano, Mastrandrea, Rossana, and Garlaschelli, Diego. 2015. Unbiased sampling of network ensembles. *New Journal of Physics*, **17**, 023052.

Squartini, Tiziano, Almog, Assaf, Caldarelli, Guido, van Lelyveld, Iman, Garlaschelli, Diego, and Cimini, Giulio. 2017a. Enhanced capital-asset pricing model for the reconstruction of bipartite financial networks. *Physical Review E*, **96**, 032315.

Squartini, Tiziano, Cimini, Giulio, Gabrielli, Andrea, and Garlaschelli, Diego. 2017b. Network reconstruction via density sampling. *Applied Network Science*, **2**(1), 3.

Squartini, Tiziano, Caldarelli, Guido, Cimini, Giulio, Gabrielli, Andrea, and Garlaschelli, Diego. 2018. Reconstruction methods for networks: The case of economic and financial systems. *Physics Reports*, **757**, 1–47.

Stone, Lewi, Simberloff, Daniel, and Artzy-Randrup, Yael. 2019. Network motifs and their origins. *PLoS Computational Biology*, **15**(4), e1006749.

Stouffer, Daniel B., Sales-Pardo, Marta, Sirer, M. Irmak, and Bascompte, Jordi. 2012. Evolutionary conservation of species' roles in food webs. *Science*, **335**(6075), 1489–1492.

Sun, Yizhou, Han, Jiawei, Aggarwal, Charu C., and Chawla, Nitesh V. 2012. When will it happen? Relationship prediction in heterogeneous information networks. Pages 663–672 of: *Proceedings of the Fifth ACM International Conference on Web Search and Data Mining*. WSDM '12. New York: ACM.

Tacchella, Andrea, Cristelli, Matthieu, Caldarelli, Guido, Gabrielli, Andrea, and Pietronero, Luciano. 2012. A new metrics for countries' fitness and products' complexity. *Scientific Reports*, **2**, 723.

Tan, Fei, Xia, Yongxiang, and Zhu, Boyao. 2014. Link prediction in complex networks: A mutual information perspective. *PLoS ONE*, **9**(9), e0107056.

Tanaka, R., Csete, M., and Doyle, J. 2005. Highly optimised global organisation of metabolic networks. *IEE Proceedings-Systems Biology*, **152**(4), 179–184.

Tong, Hanghang, Faloutsos, Christos, and Pan, Jia-Yu. 2006. Fast random walk with restart and its applications. Pages 613–622 of: *Proceedings of the Sixth*

International Conference on Data Mining. ICDM '06. Washington, DC, USA: IEEE Computer Society.

Tunç, Birkan, and Verma, Ragini. 2015. Unifying inference of meso-scale structures in networks. *PLoS ONE*, **10**(11), e0143133.

Tylenda, Tomasz, Angelova, Ralitsa, and Bedathur, Srikanta. 2009. Towards time-aware link prediction in evolving social networks. In: *Proceedings of the 3rd Workshop on Social Network Mining and Analysis*. SNA-KDD '09. New York, NY, USA: ACM.

Uchida, Hirofumi et al. 2015. *Interfirm Relationships and Trade Credit in Japan*. SpringerBriefs in Economics. Springer.

Upper, Christian. 2011. Simulation methods to assess the danger of contagion in interbank markets. *Journal of Financial Stability*, **7**(3), 111–125.

van Lidth de Jeude, Jeroen, Caldarelli, Guido, and Squartini, Tiziano. 2019a. Detecting core-periphery structures by surprise. *EPL (Europhysics Letters)*, **125**(6), 68001.

van Lidth de Jeude, Jeroen, Di Clemente, Riccardo, Caldarelli, Guido, Saracco, Fabio, and Squartini, Tiziano. 2019b. Reconstructing mesoscale network structures. *Complexity*, **2019**.

Veld, Daan in't, and van Lelyveld, Iman. 2014. Finding the core: Network structure in interbank markets. *Journal of Banking & Finance*, **49**, 27–40.

Vitali, Stefania, Glattfelder, James B, and Battiston, Stefano. 2011. The network of global corporate control. *PLoS one*, **6**(10), e25995.

Wang, Tong, He, Xing-Sheng, Zhou, Ming-Yang, and Fu, Zhong-Qian. 2017. Link prediction in evolving networks based on popularity of nodes. *Scientific Reports*, **7**, 7147.

Wang, Zuxi, Wu, Yao, Li, Qingguang, Jin, Fengdong, and Xiong, Wei. 2016. Link prediction based on hyperbolic mapping with community structure for complex networks. *Physica A: Statistical Mechanics and Its Applications*, **450**, 609–623.

Watanabe, Tsutomu, Uesugi, Iichiro, and Ono, Arito (eds.). 2015. *The Economics of Interfirm Networks*. Springer Japan.

Watts, D. J., and Strogatz, S. H. 1998. Collective dynamics of 'small-world' networks. *Nature*, **393**(6684), 440–442.

Wells, Simon. 2004. *Financial interlinkages in the United Kingdom's interbank market and the risk of contagion*. Working Paper 230. Bank of England.

Xiang, Bing-Bing, Bao, Zhong-Kui, Ma, Chuang, Zhang, Xingyi, Chen, Han-Shuang, and Zhang, Hai-Feng. 2018. A unified method of detecting core-periphery structure and community structure in networks. *Chaos: An Interdisciplinary Journal of Nonlinear Science*, **28**(1), 013122.

Xu, Zhiqiang. 2018. The minimal measurement number for low-rank matrix recovery. *Applied and Computational Harmonic Analysis*, **44**(2), 497–508.

Xu, Zhongqi, Pu, Cunlai, and Yang, Jian. 2016. Link prediction based on path entropy. *Physica A: Statistical Mechanics and Its Applications*, **456**, 294–301.

Yang, Jaewon, and Leskovec, Jure. 2014. Overlapping communities explain core–periphery organization of networks. *Proceedings of the IEEE*, **102**(12), 1892–1902.

Yang, Rong, Zhuhadar, Leyla, and Nasraoui, Olfa. 2011. Bow-tie decomposition in directed graphs. Pages 1–5 of: *14th International Conference on Information Fusion*. IEEE.

Yang, Yang, Lichtenwalter, Ryan N., and Chawla, Nitesh V. 2015. Evaluating link prediction methods. *Knowledge and Information Systems*, **45**, 751–782.

Yeger-Lotem, Esti, Sattath, Shmuel, Kashtan, Nadav, Itzkovitz, Shalev, Milo, Ron, Pinter, Ron Y., Alon, Uri, and Margalit, Hanah. 2004. Network motifs in integrated cellular networks of transcription–regulation and protein–protein interaction. *Proceedings of the National Academy of Sciences*, **101**(16), 5934–5939.

Zeng, An, and Cimini, Giulio. 2012. Removing spurious interactions in complex networks. *Physical Review E*, **85**, 036101.

Zhang, Jun, Ackerman, Mark S., and Adamic, Lada. 2007. Expertise networks in online communities: structure and algorithms. Pages 221–230 of: *Proceedings of the 16th International Conference on World Wide Web*. ACM.

Zhang, Xiao, Martin, Travis, and Newman, Mark E. J. 2015. Identification of core-periphery structure in networks. *Physical Review E*, **91**(3), 032803.

Zhao, Jing, Tao, Lin, Yu, Hong, Luo, JianHua, Cao, ZhiWei, and Li, YiXue. 2007. Bow-tie topological features of metabolic networks and the functional significance. *Chinese Science Bulletin*, **52**(8), 1036–1045.

Zhao, Jing, Miao, Lili, Yang, Jian, Fang, Haiyang, Zhang, Qian-Ming, Nie, Min, Holme, Petter, and Zhou, Tao. 2015. Prediction of links and weights in networks by reliable routes. *Scientific Reports*, **5**, 12261.

Zhou, Tao, Lü, Linyuan, and Zhang, Yi-Cheng. 2009. Predicting missing links via local information. *The European Physical Journal B*, **71**(4), 623–630.

Zhu, Boyao, and Xia, Yongxiang. 2015. An information-theoretic model for link prediction in complex networks. *Scientific Reports*, **5**, 13707.

Disclaimer

Authors' contributions: GC wrote Section 4. RM wrote Section 3. TS wrote Section 2. All authors wrote Sections 1 and 5, and reviewed and approved the manuscript.

Cambridge Elements ☰

The Structure and Dynamics of Complex Networks

Guido Caldarelli

Ca' Foscari University of Venice

Guido Caldarelli is Full Professor of Theoretical Physics at Ca' Foscari University of Venice. Guido Caldarelli received his Ph.D. from SISSA, after which he held postdoctoral positions in the Department of Physics and School of Biology, University of Manchester, and the Theory of Condensed Matter Group, University of Cambridge. He also spent some time at the University of Fribourg in Switzerland, at École Normale Supérieure in Paris, and at the University of Barcelona. His main research focus is the study of networks, mostly analysis and modelling, with applications from financial networks to social systems as in the case of disinformation. He is the author of more than 200 journal publications on the subject, and three books, and is the current President of the Complex Systems Society (2018 to 2021).

About the Series

This cutting-edge new series provides authoritative and detailed coverage of the underlying theory of complex networks, specifically their structure and dynamical properties. Each Element within the series will focus upon one of three primary topics: static networks, dynamical networks, and numerical/computing network resources.

Cambridge Elements $\equiv$

The Structure and Dynamics of Complex Networks

Printed in the United States
by Baker & Taylor Publisher Services